REVISE EDEXCEL AS
Mathematics

PRACTICE PAPERS Plus⁺

Authors: Glyn Payne and Richard Carter

Also available to support your revision:

Revise A Level Revision Planner   9781292191546

The **Revise A Level Revision Planner** helps you to plan and organise your time, step-by-step, throughout your A level revision. Use this book and wall chart to mastermind your revision.

These Practice Papers are designed to complement your revision and to help prepare you for the exams. They do not include all the content and skills needed for the complete course and have been written to help you practise what you have learned. They may not be representative of a real exam paper. Remember that the official Pearson specification and associated assessment guidance materials are the only authoritative source of information and should always be referred to for definitive guidance.

For further information, go to quals.pearson.com/alevelmathssupport

 Pearson

Published by Pearson Education Limited, 80 Strand, London, WC2R 0RL.

www.pearsonschoolsandfecolleges.co.uk

Copies of official specifications for all Pearson qualifications may be found on the website:
qualifications.pearson.com

Text and illustrations © Pearson Education Ltd 2018
Typeset and illustrated by Tech-Set Ltd, Gateshead
Produced by Project One Publishing Solutions
Cover illustration by Miriam Sturdee

The rights of Glyn Payne and Richard Carter to be identified as authors of this work have been asserted
by them in accordance with the Copyright, Designs and Patents Act 1988.

First published 2018

21 20 19 18

10 9 8 7 6 5 4 3 2 1

British Library Cataloguing in Publication Data
A catalogue record for this book is available from the British Library

ISBN 978 1 292 21327 9

Printed in Slovakia by Neografia

Notes from the publisher

1. While the publishers have made every attempt to ensure that advice on the qualification and its
assessment is accurate, the official specification and associated assessment guidance materials are the
only authoritative source of information and should always be referred to for definitive guidance.

Pearson examiners have not contributed to any sections in this resource relevant to examination papers
for which they have responsibility.

2. Pearson has robust editorial processes, including answer and fact checks, to ensure the accuracy of the
content in this publication, and every effort is made to ensure this publication is free of errors. We are,
however, only human, and occasionally errors do occur. Pearson is not liable for any misunderstandings
that arise as a result of errors in this publication, but it is our priority to ensure that the content is
accurate. If you spot an error, please do contact us at resourcescorrections@pearson.com so we can make
sure it is corrected.

Contents

Using this book   iv

Set A Paper 1 Pure Mathematics   1

Set A Paper 2 Statistics and Mechanics   17

Set B Paper 1 Pure Mathematics   27

Set B Paper 2 Statistics and Mechanics   43

Solutions

Set A Paper 1 Pure Mathematics   52

Set A Paper 2 Statistics and Mechanics   68

Set B Paper 1 Pure Mathematics   78

Set B Paper 2 Statistics and Mechanics   94

Formulae sheet   103

Statistical tables   104

Using this book

This book has been created to help you prepare for your exam by familiarising yourself with the approach of the papers and the exam-style questions. Unlike the exam, however, all of the questions have targeted hints, guidance and support in the margin to help you understand how to tackle them.

All questions also have fully worked solutions shown in the back of the book for you to refer to.

You may want to work through the papers at your own pace, to reinforce your knowledge of the topics and practise the skills you have gained throughout your course. Alternatively, you might want to practise completing a paper as if in an exam. If you do this, bear these points in mind:

- Use black ink or ball-point pen.
- Answer all questions and make sure your answers to parts of questions are clearly labelled.
- Answer the questions in the spaces provided – there may be more space than you need.
- In a real exam, you must show all your working out in order to get full credit.
- You may use a calculator in both Paper 1 and Paper 2, but it must not have functions for algebra, differentiation and integration, or have retrievable formulae.
- Diagrams are not accurately drawn, unless otherwise indicated in the question.
- The marks for each question are shown in brackets. Use this as a guide to how much time to spend on each question.

Paper 1: Pure Mathematics

- The total number of marks available for each Paper 1: Pure Mathematics is 100.
- You have 2 hours to complete Paper 1.

Paper 2: Statistics and Mechanics

- The total number of marks available for each Paper 2: Statistics and Mechanics is 60.
- You have 1 hour 15 minutes to complete Paper 2.

Paper 1: Pure Mathematics

Answer ALL questions. Write your answers in the spaces provided.

1 Solve the simultaneous equations

$$2x + y = 3$$
$$x^2 + y^2 = 18$$

(Solutions based entirely on graphical or numerical methods are not acceptable.)

(6)

....

....

....

....

....

....

....

....

....

....

....

....

....

....

....

(Total for Question 1 is 6 marks)

Revision Guide
page 9

Hint

Rearrange the first equation so that y is a function of x. Then substitute this into the second equation.

Hint

Form a quadratic equation in terms of x then solve using any technique. In this case you should be able to solve the equation by factorising.

Hint

Put your solutions for x back into the first equation to get matching solutions for y.

Hint

Don't forget to check that your values for x and y work in both equations.

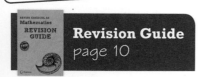

Revision Guide
page 10

Hint Q2a

Multiply out and rearrange to solve for *x*.

Hint Q2a

Remember to reverse the inequality if you multiply or divide by a negative number.

Hint Q2b

Ignore the inequality to begin with. Set the quadratic expression equal to 0 and factorise in the usual way to find two solutions for *x*. You can either use a graph to decide where the inequality is correct or substitute a value between your solutions to see if the inequality works. The coefficient of x^2 is positive, so if it does work, then you have $a < x < b$; if not, it's $x < a$ and $x > b$.

Problem solving

For part (c), combine your answers to the first two parts on a number line and check your solution works for both. Try a number of values that are spread across both inequalities to check.

2   Find the set of values of *x* for which

(a) $2(3x - 1) > 2 - 2x$

(2)

(b) $2x^2 - 11x + 12 > 0$

(4)

(c) both $2(3x - 1) > 2 - 2x$ and $2x^2 - 11x + 12 > 0$

(2)

(Total for Question 2 is 8 marks)

3 Show that $\dfrac{3 + \sqrt{24}}{3 - \sqrt{6}}$ can be written in the form $a + b\sqrt{c}$

where a, b and c are integers.

(5)

Revision Guide
page 3

Hint

You can make your working easier by simplifying any surds. Then multiply the numerator and denominator by a suitable expression to remove the surd part in the denominator.

(Total for Question 3 is 5 marks)

Revision Guide
page 13

LEARN IT!

For $y = f(x + 1)$ the transformation is inside the brackets. Move the graph one unit to the left.

LEARN IT!

For $y = f(-x)$ the transformation is inside the brackets. Reflect in the y-axis.

LEARN IT!

For $y = \dfrac{1}{2}f(x)$ the transformation is outside the brackets. Divide all the y-coordinates by 2.

4   Figure 1 shows a sketch of the curve $y = f(x)$. The curve passes through the y-axis at $(0, 7)$ and the x-axis at $(8, 0)$, has a minimum point at $(-4, 5)$ and a maximum point at $(1, 8)$.

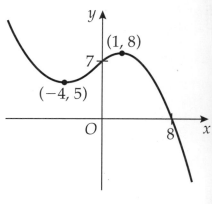

On separate diagrams, sketch the curves with the given equations, showing clearly the coordinates of any turning points, and the points at which the curve meets the coordinate axes.

Figure 1

(a) $y = f(x + 1)$

(3)

(b) $y = f(-x)$

(3)

(c) $y = \dfrac{1}{2}f(x)$

(3)

(Total for Question 4 is 9 marks)

5   Relative to a fixed origin O, the point A has position vector $(\mathbf{i} + \mathbf{j})$, the point B has position vector $(2\mathbf{i} - 3\mathbf{j})$, and the point C has position vector $(4\mathbf{i} - 11\mathbf{j})$.

Revision Guide
pages 33, 34

(a) Find the magnitude of the vector $\overrightarrow{OC}$.

(2)

Hint Q5a

Use Pythagoras.

(b) Find the angle that $\overrightarrow{OB}$ makes with the vector $\mathbf{i}$.

(2)

Hint Q5b

Show the vector $\overrightarrow{OB}$ on a sketch and make it into a right-angled triangle. Find the angle between the hypotenuse and the x-axis.

(c) Show that the points A, B and C are collinear.

(4)

Problem solving

For part (c) find the vectors $\overrightarrow{AB}$ and $\overrightarrow{BC}$ (or $\overrightarrow{AC}$ would work just as well). If they are collinear then $\overrightarrow{AB}$ and $\overrightarrow{BC}$ will be multiples of the same vector. Remember to state that the line segments have a point in common otherwise you are showing that the vectors are parallel rather than that the points are collinear.

(Total for Question 5 is 8 marks)

Revision Guide
page 40

Hint

If you see the words 'stationary points', 'maximum' or 'minimum' it's very likely that you are going to need to differentiate.

Hint

To find the stationary points, differentiate and set $\dfrac{dy}{dx} = 0$. Solve the equation to get the x-values. Substitute these back into the original equation to get the corresponding values for y.

Hint

To determine the nature of the stationary points you will need to differentiate for a second time, finding $\dfrac{d^2 y}{dx^2}$.

LEARN IT!

$\dfrac{d^2 y}{dx^2} < 0 \Rightarrow$ maximum

$\dfrac{d^2 y}{dx^2} > 0 \Rightarrow$ minimum

6 The curve C has equation

$$y = 2x^3 - 3x^2 + 7$$

Find the coordinates of the stationary points of the curve and determine the nature of each of them.

(Solutions based entirely on graphical or numerical methods are not acceptable.)

(7)

(Total for Question 6 is 7 marks)

7 Solve the equation $3^x + 9^x = 6$, $x \in \mathbb{R}$

by using a substitution to find x.

(5)

Revision Guide
page 50

Hint

$9 = 3^2$ so change 9^x into 3^{2x}. Now let $y = 3^x$ and form a quadratic that can be factorised to find y.

Hint

Once you have found y you can substitute 3^x back into the solutions and use logarithms (either base 3 or natural logarithms) to find x. In this question there is only one solution.

....

....

....

....

....

....

....

....

....

....

....

....

....

....

....

....

....

....

....

....

(Total for Question 7 is 5 marks)

Revision Guide
page 23

Hint Q8a

Find f(−3). Show every step of your substitution clearly to get both marks.

Hint Q8b

Divide f(x) by (x + 3) to get a quadratic and then factorise this quadratic to get the remaining two factors.

8 (a) Use the factor theorem to show that $(x + 3)$ is a factor of

$$4x^3 + 8x^2 - 15x - 9$$

(2)

(b) Factorise $4x^3 + 8x^2 - 15x - 9$ completely.

(4)

....

....

....

....

....

....

....

....

....

....

....

....

....

....

....

....

....

....

(Total for Question 8 is 6 marks)

9 Solve for $0 \leqslant \theta \leqslant 360°$, the equation

$$\cos 2\theta = \frac{\sqrt{3}}{2}$$

(3)

Revision Guide
page 32

Hint

As it's 2θ don't forget to write down $\cos^{-1}\left(\dfrac{\sqrt{3}}{2}\right)$ for $0 \leqslant 2\theta \leqslant 720°$. When you find solutions for 2θ, make sure you divide by 2 to give solutions for θ as required in the question.

(Total for Question 9 is 3 marks)

Revision Guide
page 45

Hint

Change $\dfrac{4}{x^2}$ to $4x^{-2}$ before you integrate.

Hint

Simplify $\sqrt{12}$ to $2\sqrt{3}$ at any point of the solution and rationalise any surds in the denominator.

10 Given that $f(x) = 3x - 2 + \dfrac{4}{x^2}$

show that $\displaystyle\int_{\sqrt{3}}^{\sqrt{12}} f(x)\,dx = \dfrac{27}{2} - \dfrac{4\sqrt{3}}{3}$

(4)

....

....

....

....

....

....

....

....

....

....

....

....

....

....

....

....

....

....

....

....

(Total for Question 10 is 4 marks)

11 (a) Find the first 3 terms, in ascending powers of x, of the binomial
       expansion of

$$(2 - 3x)^5$$

       giving each term in its simplest form.

(4)

(b) Explain how you would use your expansion to give an estimate
     for the value of 1.94^5.

(1)

Revision Guide
pages 24, 25

Hint Q11a

Be extra careful with
negative terms. Always
write every part of
each term in brackets
before you start to
work it out.

Hint Q11b

$2 - 3x = 1.94$

....

....

....

....

....

....

....

....

....

....

....

....

....

....

....

....

....

(Total for Question 11 is 5 marks)

Revision Guide
page 7

Problem solving

For any question that says 'prove that' or 'show that', always show every step of your working. You should explain what you are doing at each stage and your last line of working should be the statement that was to be proved.

Hint

Use the discriminant.

12 Prove that the roots of the equation $x^2 - 2px + p^2 - q^2 = 0$ where p and q are constants, are real.

(3)

(Total for Question 12 is 3 marks)

13

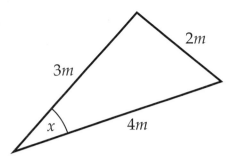

2m

3m

x   4m

Not drawn to scale

Revision Guide
page 27

A student was asked to find the value of $\cos x$ in terms of m from the triangle in the diagram.

The student's attempt is shown below:

$2m^2 = 4m^2 + 3m^2 - 2 \times 4m \times 3m\cos x$

$-5m^2 = -24m\cos x$

$\cos x = \dfrac{5m}{24}$

(a) Identify the errors made by the student.

(2)

(b) Find the correct value of $\cos x$.

(3)

(c) Use this value to find an exact value for $\sin x$.

(2)

Hint Q13a

Think about how you would work out the answer yourself, then compare this method with the sample answer given.

Hint Q13c

Draw a right-angled triangle with lengths corresponding to $\cos x$ on the adjacent and hypotenuse. Use Pythagoras to find the opposite and then use this to find $\sin x$. An exact value will include a surd for this question.

....

....

....

....

....

....

....

....

....

....

(Total for Question 13 is 7 marks)

Revision Guide
page 35

Problem solving

Use the formula for differentiation from first principles given in the formulae booklet. You need to state what happens to the expression as $h \to 0$, and write down the result you have proved.

14 Prove, from first principles, that the derivative of $\dfrac{1}{x}$ is $-\dfrac{1}{x^2}$.

(4)

(Total for Question 14 is 4 marks)

15 A hollow container in the shape of a cuboid has a base that measures $3x$ cm by x cm. The container doesn't have a top and has a height of y cm. The external surface area is 200 cm^2.

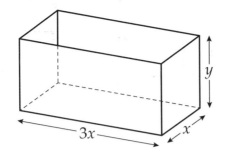

(a) Show that the volume, $V \text{ cm}^3$, of the container is given by

$$V = 75x - \frac{9x^3}{8}$$

(4)

(b) Given that x can vary, find the maximum value of V to the nearest cm^3 and justify that this value is a maximum.

(7)

Revision Guide
page 41

Modelling

For part (a), form an equation from the surface area information with y as the subject. Form an equation for V in terms of x and y and then substitute in for y.

Problem solving

In part (b), you will need to differentiate to find the maximum value. Don't forget to differentiate a second time to show that it's a maximum.

LEARN IT!

Justify means give reasons.

....

....

....

....

....

....

....

....

....

....

....

....

....

(Total for Question 15 is 11 marks)

Revision Guide
page 52

Modelling

In part (a), you will need to use ln (natural logarithms) to find k.

Hint Q16c

Think about what happens as *t* tends to infinity.

LEARN IT!

$$\frac{d(e^{kt})}{dt} = ke^{kt}$$

This is a standard differential.

16 The total sales of a new novel in thousands of books, B, t months after release are modelled by the formula

$$B = 50(1 - e^{kt})$$

In the first month it is estimated that 10 000 books are sold.

(a) Find k.

(3)

(b) How many books do the publishers expect to sell in the first 5 months?

(2)

(c) Show that, according to this model, no more than 50 000 copies will be sold in the lifetime of the book.

(2)

(d) Show that the rate of sales of the book after t months is given by $\frac{dB}{dt} = Ae^{kt}$, where A is a constant to be found to 3 significant figures.

(2)

(Total for Question 16 is 9 marks)

TOTAL FOR PAPER IS 100 MARKS

Paper 2: Statistics and Mechanics
SECTION A: STATISTICS

Revision Guide
pages 71, 72

Answer ALL questions. Write your answers in the spaces provided.

1 The Venn diagram shows the probabilities that a randomly chosen student at a college will take part in basketball and cricket.

B represents the event that a student takes part in basketball.

C represents the event that a student takes part in cricket.

x and y are probabilities.

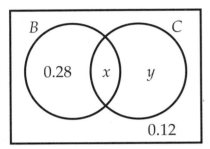

Hint Q1a

$P(C) = 1 - P(\text{not } C)$

Hint Q1b

$P(B \text{ and } C) = P(B) \times P(C)$

Hint Q1b

Use the rule and solve for x.

The events B and C are statistically independent.

(a) Write down the probability that a student plays cricket.

(1)

(b) Find the values of x and y.

(4)

....

....

....

....

....

....

....

....

....

....

(Total for Question 1 is 5 marks)

Revision Guide
pages 56, 57, 60

LEARN IT!

In a systematic sample, elements are chosen at regular intervals.

Problem solving

You need to be familiar with the large data set, though you won't need to remember exact values from it.

Hint Q2c

Use the formulae for the mean and standard deviation and give your answers to 3 s.f.

LEARN IT!

Mean $= \dfrac{\sum x}{n}$

Standard deviation

$= \sqrt{\dfrac{\sum x^2}{n} - \left(\dfrac{\sum x}{n}\right)^2}$

2   Lucy is investigating the daily minimum temperature for Leeming in the months of May to August (inclusive) 2015.

She uses the large data set for her investigation.

(a) Describe how Lucy can take a systematic sample of 30 days.

(3)

(b) From your knowledge of the large data set, explain why this might not give Lucy a sample of size 30.

(1)

The data Lucy collected are summarised as follows:

$n = 30$ $\sum x = 248$ $\sum x^2 = 2361$

(c) Calculate the mean and standard deviation of Lucy's sample.

(3)

....

....

....

....

....

....

....

....

....

....

....

....

....

....

(Total for Question 2 is 7 marks)

Revision Guide
page 73

3   A company requires applicants for jobs to take an aptitude test. The possible outcomes are: pass, fail, be retested. The probabilities for these outcomes are:

P(pass) = 0.3   P(fail) = 0.5   P(be retested) = 0.2

An applicant can be retested twice. If they do not pass after being retested twice, they are rejected.

The probabilities for the outcomes with the first retesting are:

P(pass) = 0.3   P(fail) = 0.5   P(be retested) = 0.2

The probabilities for the outcomes with the second retesting are:

P(pass) = 0.4   P(fail) = 0.6

(a)   Draw a tree diagram to illustrate these outcomes.

(3)

Hint Q3a

This will be a 3-stage tree but the 2nd and 3rd stages only follow the 'retested' branches.

(b)   Find the probability that a randomly selected applicant is accepted.

(2)

Hint Q3b

There are three ways an applicant can be successful. Work out each probability and add them together.

(c)   Three friends apply for jobs with this company. What is the probability that two of them pass at the first attempt and one of them at the final attempt?

(3)

Hint Q3c

Be careful – consider how many ways this can be done.

(Total for Question 3 is 8 marks)

19

Revision Guide
pages 75, 76, 77

Hint Q4a

You need to know when a binomial distribution can be used.

Problem solving

Use the binomial probability distribution to express the problem to be solved in terms of clearly defined mathematical statements. Finding the critical region enables you to comment on actual outcomes.

Hint Q4b

Write down the parameters you are going to use.

Hint Q4c

Be careful. This is a two-tailed test.

Hint Q4d

5% in each tail. $n = 20$, $p = 0.4$. Look for probabilities < 0.05

4   A coffee shop provides its customers with wi-fi.

The probability that a randomly selected customer uses this wi-fi is 0.4. Twelve customers are selected at random.

(a) Give two reasons why the number of customers in the sample who use wi-fi can be modelled using a binomial distribution.

(2)

(b) Find the probability that at least six of the customers in the sample used the wi-fi.

(2)

Another branch of this coffee shop also provides its customers with free wi-fi. The manager suspects that the probability of a customer using this facility may be different from 0.4. A random sample of 20 customers is selected.

(c) Write down the hypotheses that should be used to test the manager's theory.

(1)

(d) Using a 10% level of significance, find the critical region for a two-tailed test to investigate the manager's theory. You should state the probability of rejection in each tail, which should be less than 0.05.

(3)

(e) Find the actual significance level of the test based on your critical region from part (d).

(1)

One morning, the manager finds that 11 out of 20 customers are using the wi-fi.

(f) Comment on the manager's theory in light of this observation.

(1)

....

....

....

....

....

....

....

Hint Q4e

Add the probabilities from your critical region.

Hint Q4f

Look at how the value 11 relates to the critical region.

Modelling

Always interpret your answers in the context of the question. Don't just leave them as expressions involving mathematical symbols.

(Total for Question 4 is 10 marks)

Revision Guide
page 81

Hint Q5a

Label your sketch with the given information. Time goes on the horizontal axis.

Hint Q5b

The area under a velocity–time graph represents distance.

SECTION B: MECHANICS

Answer ALL questions. Write your answers in the spaces provided.

Unless otherwise indicated, whenever a numerical value of g is required, take $g = 9.8\,\text{m s}^{-2}$ and give your answer to either 2 significant figures or 3 significant figures.

5   A particle moves in a straight line. At time $t = 0$, the velocity is $u\,\text{m s}^{-1}$. The particle then accelerates with constant acceleration to $11\,\text{m s}^{-1}$ in 3 s.

The particle maintains this velocity of $11\,\text{m s}^{-1}$ for another 10 s.

It then decelerates with constant deceleration to rest in a further 2 s.

(a)   Sketch a velocity–time graph to illustrate the motion of the particle.

(3)

(b)   If the total distance travelled is 148 m, work out the value of u.

(3)

(Total for Question 5 is 6 marks)

6   A particle moves along the x-axis with velocity $v\,\mathrm{m\,s^{-1}}$.

At time t seconds, the velocity of the particle is given by

$v = 35t - 10t^2$

The positive direction is in the sense of increasing x.

Find the distance travelled between the times when the particle is instantaneously at rest.

(5)

Revision Guide
pages 91, 92

Problem solving

You are given an expression for v in terms of t but the question asks for distance travelled, with certain conditions applied, so you need to work out a strategy.

Hint

You can't find the distance (by integration) until you first work out the times when the particle is instantaneously at rest.

(Total for Question 6 is 5 marks)

Revision Guide
pages 82, 83

Hint Q7a

Use appropriate **suvat** formulae to set up two simultaneous equations in *a* and *u*.

Hint Q7b

You know *u*, *t* and *a*, and you need to find *s*.

7   A car is driven with constant acceleration, $a \, \text{m s}^{-2}$, along a straight road.

Its speed when it passes a road sign is $u \, \text{m s}^{-1}$.

In the first 3 seconds after passing the sign, the car travels 36 m.

5 seconds after passing the sign, the car has a speed of $26 \, \text{m s}^{-1}$.

(a) Write down two equations connecting *a* and *u*. Hence find the values of *a* and *u*.

(5)

(b) Find the total distance travelled by the car in the first 5 seconds after it passes the sign.

(2)

....

....

....

....

....

....

....

....

....

....

....

....

....

....

(Total for Question 7 is 7 marks)

8 A block A of mass 4 kg is held at rest on a rough horizontal table.

It is attached to one end of a light inextensible string.

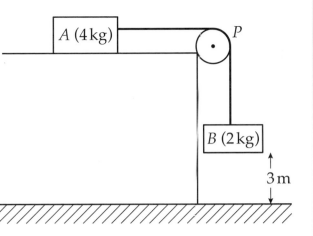

The string passes over a small smooth pulley P fixed at the edge of the table.

The other end of the string is attached to a block B of mass 2 kg, hanging freely below P and with B at a height of 3 m above the horizontal floor.

The system is released from rest with the string taut.

The resistance to the motion of A from the rough table is modelled as having constant magnitude 14.8 N.

A and B are modelled as particles.

(a) Show that the acceleration of B is $0.8\,\text{m s}^{-2}$.

(3)

(b) Calculate the tension in the string.

(2)

(c) Calculate the speed with which B hits the floor.

(2)

Block A never reaches the pulley at the edge of the table.

(d) After B has hit the floor, calculate the further distance A travels before coming to rest.

(4)

(e) State how, in your calculations, you have used the fact that the string is inextensible.

(1)

Hint Q8a

Put in all the forces acting on the blocks, then use Newton's 3rd law for both particles and solve for a.

Problem solving

Label your diagram with all the forces acting on the two blocks. The pulley is smooth, so the tension in the string is the same on each side of the pulley. Once B has hit the floor you will need to re-visit the forces acting on A.

Modelling

Initially the blocks are connected by a light inextensible string and they are modelled as particles.

Hint Q8b

Substitute to find T.

Hint Q8c

A **suvat** equation is needed here.

Aiming higher

In part (d), you must find the deceleration of A before you can use the appropriate **suvat** equation. Think about the forces acting on A after B hits the floor and the string becomes slack.

Hint Q8e

Think back to the original motion of the blocks.

(Total for Question 8 is 12 marks)

TOTAL FOR PAPER IS 60 MARKS

Paper 1: Pure Mathematics

Answer ALL questions. Write your answers in the spaces provided.

1   (a)   Express 27^{4k+2} in the form 9^m, giving m in terms of k.

(3)

(b)   Find the value of n in the equation

$$\frac{8^{2n+1}}{4^n} = 32$$

(3)

Revision Guide
page 1

Hint Q1a

Start by writing 27 as a power of 3, then write 3 as a power of 9.

Hint Q1b

Change 8, 4 and 32 into powers of 2.

....

....

....

....

....

....

....

....

....

....

....

....

....

....

....

....

....

....

....

(Total for Question 1 is 6 marks)

Revision Guide
page 4

Hint

Start by dividing the equation by a. Now complete the square and rearrange for x.

2 $ax^2 + bx + c = 0$.

Prove, by completing the square, that

$$x = \frac{-b \pm \sqrt{b^2 - 4ac}}{2a}$$

(6)

(Total for Question 2 is 6 marks)

3   $f(x) = (2 + kx)^7$, where k is a constant.

Given that the coefficient of x^3 in the binomial expansion of
$f(x)$ is 70, find the value of k.

(3)

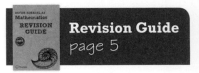

Revision Guide
page 5

Hint

Use the *nCr* button on your calculator to find the coefficient of the term then multiply this by 2^4 and $(kx)^3$. This will give the x^3 term which can be equated to 70, then solve to find k.

(Total for Question 3 is 3 marks)

Revision Guide
pages 20, 22

Hint Q4a

Substitute $y = 2x + 11$ into the circle equation and show there is only one solution. If there is one repeated root, then the line must touch the circle at exactly one point, so it must be a tangent.

Hint Q4b

It might help to sketch the circle and the line. The first step will be to find the point of contact. Use the fact that the tangent to a circle is perpendicular to a radius at the point of contact.

4   (a)   Show that the line $y = 2x + 11$ is a tangent to the circle

$$x^2 + y^2 - 6x - 4y = 32$$

(5)

(b)   Find the equation of the diameter of the circle that passes through the point of contact of the tangent with the circle.

(3)

....

....

....

....

....

....

....

....

....

....

....

....

....

....

....

....

....

....

....

(Total for Question 4 is 8 marks)

5   Prove that

$$\frac{1 - \sin\theta}{1 + \sin\theta} \equiv \left(\frac{1}{\cos\theta} - \tan\theta\right)^2$$

(5)

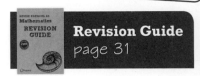

Revision Guide
page 31

LEARN IT!

You will need to use the trigonometrical identities $\sin^2\theta + \cos^2\theta \equiv 1$ and $\frac{\sin\theta}{\cos\theta} \equiv \tan\theta$.

Hint

Multiply the left-hand side by $\frac{1 - \sin\theta}{1 - \sin\theta}$
Then work through from the left-hand side to the right-hand side.

(Total for Question 5 is 5 marks)

Revision Guide
page 15

Hint Q6a

The reciprocal graph is a transformation of the graph $y = \dfrac{2}{x}$

Hint Q6b

Write $\dfrac{2}{x + 1} = 2x + 3$ and rearrange to form a quadratic. Telling you that you need to round to 2 d.p. is a giveaway that you need to use the quadratic formula.

6 (a) On the same axes sketch the graphs

$$y = \frac{2}{x + 1}$$
$$y = 2x + 3$$

Show where the graphs cross the axes and any asymptotes.

(4)

(b) Find the coordinates of the points of intersection of the two graphs. Round your answers to 2 d.p.

(3)

(Total for Question 6 is 7 marks)

Revision Guide
page 39

7 (a) Given that $f(x) = \dfrac{(2x + 3)(x - 5)}{x}$, $x \neq 0$, find $f'(x)$.

(3)

(b) Show that $f(x)$ is increasing for the interval $1 < x < 2$

(2)

Hint Q7a

Expand the brackets and simplify the expression first.

LEARN IT!

A function $f(x)$ is increasing if $f'(x) \geq 0$ for all values of x in the given interval.

(Total for Question 7 is 5 marks)

Revision Guide
page 2

Hint

Remove a common factor then use the difference of two squares, done twice.

8   Factorise fully

$$2x^5 - 32x$$

(3)

8   Factorise fully

$$2x^5 - 32x$$

(Total for Question 8 is 3 marks)

9 Figures 1 and 2 show the curve with equation $y = 3x^2 + 2x - 1$

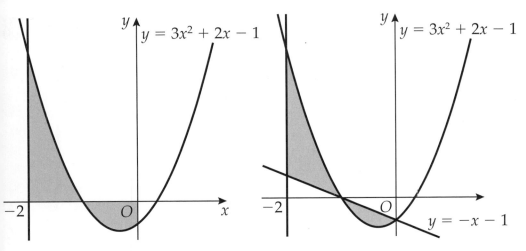

Figure 1   **Figure 2**

Revision Guide
pages 45, 46

Hint Q9a

Factorise to find the points of intersection with the *x*-axis, then sketch the quadratic graph.

Problem solving

For part (b), add $x + y = -1$ to your sketch graph and work out points of intersection so you can see the area required.

(a) Find the total shaded area shown in Figure 1, enclosed by the curve, the *x*-axis and the lines $x = -2$ and $x = 0$.

(4)

(b) Hence or otherwise find the shaded area shown in Figure 2 enclosed by the curve, the line $x + y = -1$ and the lines $x = -2$ and $x = 0$.

(3)

....

....

....

....

....

....

....

....

....

....

....

....

(Total for Question 9 is 7 marks)

Revision Guide
page 26

Problem solving

You can write a general multiple of 4 as 4k, where k is a positive integer. Find a general expression for the sum of three consecutive multiples of 4 and show that it has 12 as a factor.

10 Prove that the sum of three consecutive multiples of 4 is always a multiple of 12.

(3)

(Total for Question 10 is 3 marks)

11 (a) Find the equation of the straight line that passes through the points A $(-2, 7)$ and B $(3, 5)$, in the form $ax + by + c = 0$.

(3)

(b) Find the equation of the perpendicular bisector of AB.

(4)

(c) Point C lies on the perpendicular bisector of AB and is vertically above point B. Find the area of the triangle ABC.

(5)

Revision Guide
pages 17, 18, 19

Hint Q11a

Use gradient $= \dfrac{y_2 - y_1}{x_2 - x_1}$
and then $y - y_1 = m(x - x_1)$
to find the equation of
the straight line, then
rearrange into the form
requested.

Hint Q11b

You will need the
perpendicular gradient
and the coordinates of
the midpoint of AB.

Problem solving

In part (c), point C
must have the same
x coordinate as B.
Substitute this value
in the equation of the
perpendicular to find
the coordinates of C.
Use the point $P(-2, 5)$
to make a trapezium
$APBC$ and a triangle
APB. Find the area
of the trapezium and
subtract the area of
the triangle.

(Total for Question 11 is 12 marks)

Revision Guide
pages 42, 43

Hint Q12a

After integrating, substitute the given values to find the constant of integration.

Hint Q12b

Substitute $x = -1$ into f(x) using the constant of integration found in part (a).

12 $y = \int \left(2x^3 - 4 + \dfrac{1}{x^2} \right) dx$

(a) Given that $y = 0$ when $x = 2$, find y as a function of x.

(3)

(b) The point $P(-1, t)$ lies on the curve with equation $y = $ f(x). Find the value of t.

(2)

(Total for Question 12 is 5 marks)

13 The value of Andy's car can be modelled by the formula

$$V = 18\,600e^{-0.23t} + k$$

where £V is the value of the car, t is the age in years and k is a positive constant.

(a) The value of the car when new was £19 550. Find the value of k.

(1)

(b) Find the value of the car after 3 years, to the nearest £.

(2)

(c) Find the rate of decrease in the value of the car in £ per year to the nearest £ at the instant when the car is 5 years old.

(3)

(d) Sketch the graph of V against t.

(2)

(e) Interpret the meaning of the value of k.

(1)

Hint Q13a

$t = 0$ when the car is new.

Hint Q13b

After 3 years, $t = 3$. Use the value of k found in part (a).

Hint Q13c

Differentiate with respect to t.

Hint Q13d

Plot V on the vertical axis and t on the horizontal. Label V when $t = 0$ on the graph.

Modelling

In part (e), think about what happens to V as t increases.

(Total for Question 13 is 9 marks)

Revision Guide
pages 48, 50

Hint Q14a

Take logs of both sides. Don't round your answers until the very end.

LEARN IT!

$\log(a^b) = b\log a$

$\log(a) - \log(b) = \log\dfrac{a}{b}$

14 (a) Solve $5^{2x} = 8^{x-1}$ giving your answer to 3 significant figures.

(3)

(b) Solve $\log_3(x + 1) - \log_3(x) = 2$.

(3)

(Total for Question 14 is 6 marks)

15 A parallelogram $ABCD$ has $\overrightarrow{AB} = \begin{pmatrix} 2 \\ 5 \end{pmatrix}$ and $\overrightarrow{AD} = \begin{pmatrix} 4 \\ 1 \end{pmatrix}$.

(a) Find $\left| \overrightarrow{BD} \right|$.

(2)

(b) Find angle $B\hat{A}D$.

(3)

(c) Find the area of parallelogram $ABCD$.

(2)

Revision Guide
pages 27, 33, 34

LEARN IT!

$\overrightarrow{BD} = \overrightarrow{BA} + \overrightarrow{AD}$
$\phantom{\overrightarrow{BD}} = \overrightarrow{AD} - \overrightarrow{AB}$

Hint Q15b

There are other ways of doing this but finding the lengths of the three sides and using the cosine rule is the most straightforward.

LEARN IT!

Area of a triangle is $\frac{1}{2}|a||b|\sin C$ so area of a parallelogram is $|a||b|\sin C$.

(Total for Question 15 is 7 marks)

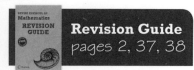

Revision Guide
pages 2, 37, 38

Problem solving

The wording makes this question look harder than it is. Sometimes you will need to pick out key words. You have seen the word 'tangent' so differentiate $y = ax^3 + x$ and then substitute $x = 1$. Then use the fact that you know the gradient of the tangent from its equation.

Hint Q16b

Use the value of a you found in part (a).

Aiming higher

In part (c), equate the two functions of x and rearrange to form a cubic equation. Use the knowledge that where the tangent touches the curve there are two solutions at the same place (so you know two of the solutions already). This will help you factorise the cubic expression.

16 The line L, defined by $y = 7x + k$, is a tangent to the curve C, defined by $y = ax^3 + x$, at the point where $x = 1$.
a and k are both constants.

(a) Find the value of a.

(2)

(b) Find the value of k.

(2)

(c) Find the coordinates of the point where $y = 7x + k$ crosses the curve $y = ax^3 + x$.

(4)

(Total for Question 16 is 8 marks)

TOTAL FOR PAPER IS 100 MARKS

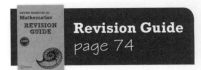

Revision Guide
page 74

Paper 2: Statistics and Mechanics
SECTION A: STATISTICS

Answer ALL questions. Write your answers in the spaces provided.

1   The discrete random variable Y can take only values 1, 2, 3, 4 and 5.

Y has probability function

$$P(Y = y) = \begin{cases} k(4 - y)^2, & y = 1, 2, 3 \\ k(3y - 8), & y = 4, 5 \end{cases}$$

where k is a constant.

(a)   Find the value of k and construct a table giving the probability distribution of Y.

(4)

(b)   Find $P(Y \geqslant 4)$

(1)

Hint Q1a

Use the probability function for values of Y from 1 to 5. Remember that the sum of all probabilities = 1.

Hint Q1b

Use the appropriate values from your probability distribution.

(Total for Question 1 is 5 marks)

Revision Guide
pages 58, 63

Hint Q2a

$n = 13$ so, for example, $\frac{n}{4} = 3.25$.

This is rounded up to 4, so Q_1 is the 4th observation from an ordered list.

Hint Q2b

Identify any outliers, deduce the whiskers, then draw the box plot using your Q_1, Q_2 and Q_3 values.

Aiming higher

The whiskers can be drawn either at the outlier boundaries, $Q_1 - 1.5 \times$ IQR and $Q_3 + 1.5 \times$ IQR, or at the smallest and largest data points which are not outliers.

2   Thirteen students sat a test, marked out of 50. These are their scores:

   33, 25, 36, 40, 14, 49, 31, 17, 31, 44, 35, 28, 34

(a) Write down the values of Q_1, Q_2 and Q_3 for this data.

(3)

A value that is more than 1.5 times the interquartile range (IQR) above Q_3 or more than 1.5 times the IQR below Q_1 is called an outlier.

(b) Draw a box plot for this data.

(4)

(Total for Question 2 is 7 marks)

3 A farmer collects data on annual rainfall, r cm, and the annual yield of broccoli, b tonnes per acre.

The table shows the results for 10 consecutive years.

r	75	79	78	84	74	76	80	85	77	81
b	4.5	4.8	4.7	5.1	4.6	4.6	4.9	5.2	4.7	4.8

The equation of the regression line of b on r is $b = 1.6 + 0.04r$.

(a) Give an interpretation of the value 0.04 in the regression equation.

(1)

(b) Comment on the reliability of using this regression equation to estimate the yield of broccoli in a year when the annual rainfall is

(i) 70 cm   (ii) 82 cm

(2)

(c) Explain why the regression equation is not suitable to estimate the annual rainfall in a year when the yield of broccoli is 4.5 tonnes per acre.

(1)

....

....

....

....

....

....

....

....

....

....

....

....

(Total for Question 3 is 4 marks)

Revision Guide
pages 62, 68, 69

Hint Q3a

Relate your interpretation to the data, so write about amount of rainfall with respect to yield per acre.

Hint Q3b

Look at the range of the original data.

LEARN IT!

The response in part (c) here is a standard one. Consider independent and dependent variables.

Revision Guide
page 59

Hint

Times are given to the nearest minute, so 5–9 really means 4.5–9.5

Hint

Work out in which class intervals the 10th and 90th percentiles lie, then work out the positions within those intervals using linear interpolation.

Hint

Watch out! The class boundaries are not whole numbers.

4 In a survey, 120 shoppers in a supermarket were asked how many minutes, to the nearest minute, they had been in the store. The results are summarised in the table.

Number of minutes	Number of shoppers
1 – 4	5
5 – 9	20
10 – 19	28
20 – 29	51
30 – 59	16
Total	**120**

Use linear interpolation to estimate the 10% to 90% interpercentile range of this data.

(5)

....

....

....

....

....

....

....

....

....

....

....

....

....

....

....

(Total for Question 4 is 5 marks)

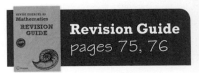

5 All cars more than three years old have to undergo an annual MOT test. 20% of cars fail the MOT test because of problems with their lights. This is the most common cause of failure nationally. A group of 5 randomly selected cars undergo an MOT test.

(a) Give a reason why the binomial distribution is suitable for modelling the number of cars failing the MOT test due to a problem with their lights.

(1)

(b) Find the probability that exactly 2 of them fail the lights test.

(2)

A car hire firm has 15 cars taking their MOT test. Six of them fail the lights test.

(c) Using a 5% level of significance, find whether there is evidence to support the suggestion that cars from this firm fail the lights test more than the national average. You should state clearly the hypotheses that should be used.

(6)

LEARN IT!

You need to know when a binomial distribution can be used.

Hint Q5b

You need to use the number of ways of selecting 2 outcomes from 5 trials, with $p = 0.2$

Hint Q5c

When carrying out any hypothesis test, you should always write down your hypotheses clearly. This is a one-tailed test so your alternative hypothesis will be in the form

$H_1: p > \ldots$ or $H_1: p < \ldots$

Modelling

Always interpret your answers in the context of the question.

....

....

....

....

....

....

....

....

....

....

....

....

(Total for Question 5 is 9 marks)

Revision Guide
page 84

Hint Q6a

For vertical motion under gravity, always choose a positive direction (up or down). In this case, if you choose upwards as positive, then $a = -9.8\,\text{ms}^{-2}$ since gravity acts downwards.

Problem solving

If the stone starts at a height of h, then it will hit the ground when $s = -h$.

LEARN IT!

Standard modelling assumptions apply here.
$s = ut + \frac{1}{2}at^2$

Modelling

In part (b), you are asked for modelling assumptions. Think about the standard conditions that apply in questions involving vertical motion under gravity.

SECTION B: MECHANICS

Answer ALL questions. Write your answers in the spaces provided.

Unless otherwise indicated, whenever a numerical value of g is required, take $g = 9.8\,\text{ms}^{-2}$ and give your answer to either 2 significant figures or 3 significant figures.

6 A stone is thrown vertically upwards with speed $15\,\text{ms}^{-1}$ from a point h metres above the ground.

The stone hits the ground 4 seconds later.

(a) Find the value of h.

(3)

(b) State two modelling assumptions made when calculating your answer.

(2)

(Total for Question 6 is 5 marks)

7 A lift of mass 220 kg is being lowered into a shaft by a vertical cable attached to the top of the lift.

The cable exerts an upward force of 2150 N on the lift.

A container of mass m kg rests on the floor of the lift.

There is a constant upward resistance of 160 N on the lift and the normal reaction between the container and the lift is of magnitude 480 N.

The lift descends with constant acceleration, a m s^{-2}.

(a) Find the acceleration of the lift.

(4)

(b) Find the mass of the container.

(3)

(Total for Question 7 is 7 marks)

Revision Guide
page 85

Hint Q7a

Use Newton's 3rd law. You can apply it to the lift, the container or the whole system. Choose whichever is the most appropriate.

Problem solving

Draw a diagram showing all the forces acting.

Hint Q7a

Put in all the forces acting on whatever part of the system you choose. Take care with signs – take downwards as positive.

Modelling

The system is such that all parts are moving in the same straight line, so you can treat the whole system as a single particle or consider parts of the system separately. The container and the lift remain in contact, so they exert equal and opposite forces on each other.

Hint Q7b

Another application of Newton's 3rd law is needed here.

Revision Guide
page 86

Hint Q8a

The particle is in equilibrium so the resultant of the three forces must be the zero vector. Set up two simultaneous equations in *a* and *b*.

Problem solving

Forces are vectors, so in part (a) use the equilibrium condition to work out the values of *a* and *b*.

Hint Q8b

Find the resultant of $\mathbf{F}_1$ and $\mathbf{F}_3$. Then find its magnitude and direction. Use $F = ma$ to find the acceleration of the particle.

Problem solving

Because the particle was in equilibrium, it is accelerating from rest, so its acceleration will act in the same direction as the resultant force.

8   Three forces, $\mathbf{F}_1$, $\mathbf{F}_2$ and $\mathbf{F}_3$ act on a particle of mass 5 kg. The forces are given as:

$$\mathbf{F}_1 = 2a\mathbf{i} - 5b\mathbf{j}$$

$$\mathbf{F}_2 = -4b\mathbf{i} + 3a\mathbf{j}$$

$$\mathbf{F}_3 = 2\mathbf{i} - 2\mathbf{j}$$

where *a* and *b* are constants.

The particle is in equilibrium.

(a) Work out the values of *a* and *b*.

(3)

The force $\mathbf{F}_2$ is removed.

(b) Find the magnitude and bearing of the resulting acceleration of the particle.

(6)

(Total for Question 8 is 9 marks)

Revision Guide
pages 91, 92

9   A rabbit leaves its burrow at time $t = 0$ and runs alongside a straight fence before returning to its burrow.

The rabbit is modelled as a particle moving in a straight line.

The distance, s metres, of the rabbit from its burrow at time t seconds is given by

$$s = \frac{1}{5}(t^4 - 16t^3 + 64t^2)$$

where $0 \leqslant t \leqslant 8$

(a) Explain the restriction $0 \leqslant t \leqslant 8$

(3)

Hint Q9a

Factorise and interpret the equation in the context of the problem.

Modelling

The rabbit is modelled as a particle moving in a straight line with its distance from its burrow, s, given as a function of time, t.

(b) Show that the rabbit is initially at rest and find its distance from the burrow when it next comes to instantaneous rest.

(6)

Hint Q9b

Instantaneous rest means $v = 0$. Use calculus to find the relevant value of t then use it to calculate the distance.

Problem solving

You will need calculus to connect expressions for s and v and consider times when the rabbit is instantaneously at rest.

....

....

....

....

....

....

....

....

....

....

....

....

....

(Total for Question 9 is 9 marks)

TOTAL FOR PAPER IS 60 MARKS

Paper 1: Pure Mathematics

Answer ALL questions. Write your answers in the spaces provided.

Revision Guide
page 9

1 Solve the simultaneous equations

$$2x + y = 3 \qquad \text{①}$$
$$x^2 + y^2 = 18 \qquad \text{②}$$

(Solutions based entirely on graphical or numerical methods are not acceptable.)

(6)

Hint

Rearrange the first equation so that y is a function of x. Then substitute this into the second equation.

From ① $\qquad y = 3 - 2x$ ✔

Substitute for y into ②

$$x^2 + (3 - 2x)^2 = 18 \quad ✔$$

$$x^2 + (9 - 12x + 4x^2) = 18$$

$$5x^2 - 12x + 9 = 18 \quad ✔$$

$$5x^2 - 12x - 9 = 0$$

$$(5x + 3)(x - 3) = 0 \quad ✔$$

$$x = 3 \text{ or } x = -\frac{3}{5} \quad ✔$$

When $x = 3$, $y = (3 - 2 \times 3) = -3$

When $x = -\frac{3}{5}$, $y = \left(3 - 2 \times -\frac{3}{5}\right) = \frac{21}{5}$ ✔

Hint

Form a quadratic equation in terms of x then solve using any technique. In this case you should be able to solve the equation by factorising.

Hint

Put your solutions for x back into the first equation to get matching solutions for y.

Hint

Don't forget to check that your values for x and y work in both equations.

(Total for Question 1 is 6 marks)

1

Revision Guide
page 10

Hint Q2a

Multiply out and rearrange to solve for x.

Hint Q2a

Remember to reverse the inequality if you multiply or divide by a negative number.

Hint Q2b

Ignore the inequality to begin with. Set the quadratic expression equal to 0 and factorise in the usual way to find two solutions for x. You can either use a graph to decide where the inequality is correct or substitute a value between your solutions to see if the inequality works. The coefficient of x^2 is positive, so if it does work, then you have $a < x < b$; if not, it's $x < a$ and $x > b$.

Problem solving

For part (c), combine your answers to the first two parts on a number line and check your solution works for both. Try a number of values that are spread across both inequalities to check.

---

2 Find the set of values of x for which

(a) $2(3x - 1) > 2 - 2x$

(2)

(b) $2x^2 - 11x + 12 > 0$

(4)

(c) both $2(3x - 1) > 2 - 2x$ and $2x^2 - 11x + 12 > 0$

(2)

(a) $2(3x - 1) > 2 - 2x$

$6x - 2 > 2 - 2x$

$8x > 4$ ✔

$x > \dfrac{1}{2}$ ✔

(b)

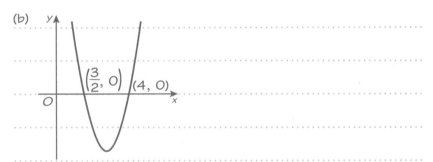

$2x^2 - 11x + 12 = 0$

$(2x - 3)(x - 4) = 0$ ✔

$x = \dfrac{3}{2}$ or $x = 4$ ✔

3 is between $\dfrac{3}{2}$ and 4 so substitute $x = 3$ into f(x).

When $x = 3$, f(x) $= 2(3)^2 - 11(3) + 12 = -3$

so f(x) < 0 and 3 doesn't work. ✔

So $x < \dfrac{3}{2}$ or $x > 4$ ✔

(c)

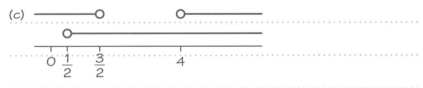

$\dfrac{1}{2} < x < \dfrac{3}{2}$ or $x > 4$ ✔✔

(Total for Question 2 is 8 marks)

2

3 Show that $\dfrac{3 + \sqrt{24}}{3 - \sqrt{6}}$ can be written in the form $a + b\sqrt{c}$

where a, b and c are integers.

(5)

Revision Guide
page 3

Hint

You can make your working easier by simplifying any surds. Then multiply the numerator and denominator by a suitable expression to remove the surd part in the denominator.

$3 + \sqrt{24} = 3 + 2\sqrt{6}$ ✓

$\dfrac{3 + 2\sqrt{6}}{3 - \sqrt{6}} = \dfrac{3 + 2\sqrt{6}}{3 - \sqrt{6}} \times \dfrac{3 + \sqrt{6}}{3 + \sqrt{6}}$ ✓

$= \dfrac{9 + 9\sqrt{6} + 12}{9 - 6}$ ✓

$= \dfrac{21 + 9\sqrt{6}}{3}$ ✓

$= 7 + 3\sqrt{6}$ ✓

$a = 7$, $b = 3$ and $c = 6$

(Total for Question 3 is 5 marks)

3

Revision Guide
page 13

LEARN IT!

For $y = f(x + 1)$ the transformation is inside the brackets. Move the graph one unit to the left.

LEARN IT!

For $y = f(-x)$ the transformation is inside the brackets. Reflect in the y-axis.

LEARN IT!

For $y = \frac{1}{2}f(x)$ the transformation is outside the brackets. Divide all the y-coordinates by 2.

4 Figure 1 shows a sketch of the curve $y = f(x)$. The curve passes through the y-axis at $(0, 7)$ and the x-axis at $(8, 0)$, has a minimum point at $(-4, 5)$ and a maximum point at $(1, 8)$.

On separate diagrams, sketch the curves with the given equations, showing clearly the coordinates of any turning points, and the points at which the curve meets the coordinate axes.

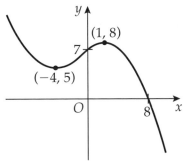

Figure 1

(a) $y = f(x + 1)$

(3)

(b) $y = f(-x)$

(3)

(c) $y = \frac{1}{2}f(x)$

(3)

(a) 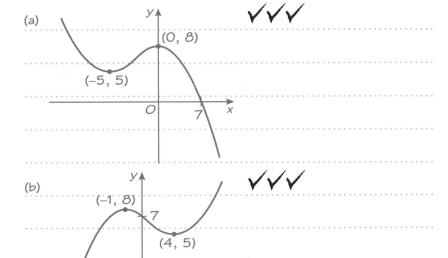 ✔✔✔

(b) ✔✔✔

(c) 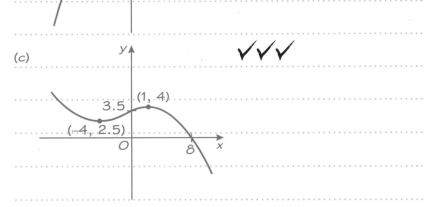 ✔✔✔

(Total for Question 4 is 9 marks)

4

5   Relative to a fixed origin O, the point A has position vector $(i + j)$, the point B has position vector $(2i - 3j)$, and the point C has position vector $(4i - 11j)$.

(a) Find the magnitude of the vector $\overrightarrow{OC}$.

(2)

(b) Find the angle that $\overrightarrow{OB}$ makes with the vector $\mathbf{i}$.

(2)

(c) Show that the points A, B and C are collinear.

(4)

Revision Guide
pages 33, 34

Hint Q5a

Use Pythagoras.

Hint Q5b

Show the vector $\overrightarrow{OB}$ on a sketch and make it into a right-angled triangle. Find the angle between the hypotenuse and the x-axis.

(a) $\left|\overrightarrow{OC}\right| = \sqrt{4^2 + 11^2}$ ✔

$= \sqrt{137} \ [= 11.7]$ ✔

(b)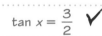

$\tan x = \dfrac{3}{2}$ ✔

$x = 56.3°$ ✔

(c) $\overrightarrow{AB} = \overrightarrow{OB} - \overrightarrow{OA} = i - 4j$ ✔

$\overrightarrow{BC} = \overrightarrow{OC} - \overrightarrow{OB} = 2i - 8j$ ✔

$\overrightarrow{BC} = 2\overrightarrow{AB}$ hence $\overrightarrow{BC}$ and $\overrightarrow{AB}$ are parallel ✔

So $\overrightarrow{AB}$ is parallel to $\overrightarrow{BC}$, and B lies on both line segments,

so A, B, and C are collinear. ✔

Problem solving

For part (c) find the vectors $\overrightarrow{AB}$ and $\overrightarrow{BC}$ (or $\overrightarrow{AC}$ would work just as well). If they are collinear then $\overrightarrow{AB}$ and $\overrightarrow{BC}$ will be multiples of the same vector. Remember to state that the line segments have a point in common otherwise you are showing that the vectors are parallel rather than that the points are collinear.

(Total for Question 5 is 8 marks)

5

Revision Guide
page 40

Hint

If you see the words 'stationary points', 'maximum' or 'minimum' it's very likely that you are going to need to differentiate.

Hint

To find the stationary points, differentiate and set $\dfrac{dy}{dx} = 0$. Solve the equation to get the x-values. Substitute these back into the original equation to get the corresponding values for y.

Hint

To determine the nature of the stationary points you will need to differentiate for a second time, finding $\dfrac{d^2 y}{dx^2}$.

LEARN IT!

$\dfrac{d^2 y}{dx^2} < 0 \Rightarrow$ maximum

$\dfrac{d^2 y}{dx^2} > 0 \Rightarrow$ minimum

6   The curve C has equation

$$y = 2x^3 - 3x^2 + 7$$

Find the coordinates of the stationary points of the curve and determine the nature of each of them.

(Solutions based entirely on graphical or numerical methods are not acceptable.)

(7)

$\dfrac{dy}{dx} = 6x^2 - 6x$ ✓

$\dfrac{dy}{dx} = 0 \Rightarrow 6x^2 - 6x = 0$ ✓

$\qquad\qquad x^2 - x = 0$

$\qquad\qquad x(x - 1) = 0$ ✓

$\qquad\qquad\qquad x = 0 \ \text{ or } \ x = 1$

When $x = 0$, $y = 2(0)^3 - 3(0)^2 + 7 = 7$

When $x = 1$, $y = 2(1)^3 - 3(1)^2 + 7 = 6$ ✓

So the stationary points are at $(0, 7)$ and $(1, 6)$.

$\dfrac{d^2 y}{dx^2} = 12x - 6$ ✓

When $x = 0$, $\dfrac{d^2 y}{dx^2} = -6$ so $(0, 7)$ is a local maximum ✓

When $x = 1$, $\dfrac{d^2 y}{dx^2} = 6$ so $(1, 6)$ is a local minimum ✓

(Total for Question 6 is 7 marks)

6

7   Solve the equation $3^x + 9^x = 6$, $x \in \mathbb{R}$

by using a substitution to find x.

(5)

Revision Guide
page 50

$3^x + (3^2)^x = 6$

$3^x + (3^x)^2 = 6$

$3^x + 3^{2x} = 6$ ✔

Let $y = 3^x \Rightarrow y + y^2 = 6$

     $y^2 + y - 6 = 0$ ✔

     $(y + 3)(y - 2) = 0$

        $y = -3$ or $y = 2$ ✔

$3^x = -3$ has no solution so not valid.   ✔

   $3^x = 2$

$\ln 3^x = \ln 2$

$x \ln 3 = \ln 2$

     $x = \dfrac{\ln 2}{\ln 3} = 0.631$ (3 s.f.)   ✔

Hint

$9 = 3^2$ so change 9^x into 3^{2x}. Now let $y = 3^x$ and form a quadratic that can be factorised to find y.

Hint

Once you have found y you can substitute 3^x back into the solutions and use logarithms (either base 3 or natural logarithms) to find x. In this question there is only one solution.

(Total for Question 7 is 5 marks)

7

58

Revision Guide
page 23

Hint Q8a

Find f(−3). Show every step of your substitution clearly to get both marks.

Hint Q8b

Divide f(x) by (x + 3) to get a quadratic and then factorise this quadratic to get the remaining two factors.

8 (a) Use the factor theorem to show that $(x + 3)$ is a factor of

$$4x^3 + 8x^2 - 15x - 9$$

(2)

(b) Factorise $4x^3 + 8x^2 - 15x - 9$ completely.

(4)

(a) $f(-3) = 4(-3)^3 + 8(-3)^2 - 15(-3) - 9$

$= 4 \times (-27) + 8 \times 9 - 15 \times (-3) - 9$ ✔

$= -108 + 72 + 45 - 9$

$= 0$ ✔

Since $f(-3) = 0$, $(x + 3)$ is a factor.

(b)
$$
\begin{array}{r}
4x^2 - 4x - 3 \\
x + 3 \overline{)\, 4x^3 + 8x^2 - 15x - 9} \\
\underline{4x^3 + 12x^2} \\
-4x^2 - 15x \\
\underline{-4x^2 - 12x} \\
-3x - 9 \\
\underline{-3x - 9} \\
0
\end{array}
$$
✔ ✔

$4x^2 - 4x - 3 = (2x + 1)(2x - 3)$ ✔

So $4x^3 + 8x^2 - 15x - 9 = (2x + 1)(2x - 3)(x + 3)$ ✔

Alternative solution

Find the quadratic factor by inspection.

$4x^3 + 8x^2 - 15x - 9 = (x + 3)(4x^2 + kx - 3)$

Equate coefficients of x^2

$8x^2 = 12x^2 + kx^2$

$k = -4$

(Total for Question 8 is 6 marks)

8

9 Solve for $0 \leqslant \theta \leqslant 360°$, the equation

$$\cos 2\theta = \frac{\sqrt{3}}{2}$$

(3)

Revision Guide
page 32

Hint

As it's 2θ don't forget to write down $\cos^{-1}\left(\dfrac{\sqrt{3}}{2}\right)$ for $0 \leqslant 2\theta \leqslant 720°$. When you find solutions for 2θ, make sure you divide by 2 to give solutions for θ as required in the question.

$\cos 2\theta \dfrac{\sqrt{3}}{2}$

$0 \leqslant \theta \leqslant 360°$

so

$0 \leqslant 2\theta \leqslant 720°$

$2\theta = \cos^{-1} \dfrac{\sqrt{3}}{2}$ ✓

$2\theta = 30°,\ 330°,\ 390°,\ 690°$ ✓

$\theta = 15°,\ 165°,\ 195°,\ 345°$ ✓

(Total for Question 9 is 3 marks)

9

60

Revision Guide
page 45

Hint

Change $\dfrac{4}{x^2}$ to $4x^{-2}$ before you integrate.

Hint

Simplify $\sqrt{12}$ to $2\sqrt{3}$ at any point of the solution and rationalise any surds in the denominator.

10 Given that $f(x) = 3x - 2 + \dfrac{4}{x^2}$

show that $\displaystyle\int_{\sqrt{3}}^{\sqrt{12}} f(x)\,dx = \dfrac{27}{2} - \dfrac{4\sqrt{3}}{3}$

(4)

$f(x) = 3x - 2 + \dfrac{4}{x^2}$

$\qquad = 3x - 2 + 4x^{-2}$ ✔

$\displaystyle\int_{\sqrt{3}}^{\sqrt{12}} (3x - 2 + 4x^{-2})\,dx = \left[\dfrac{3x^2}{2} - 2x + 4x^{-1}\right]_{\sqrt{3}}^{\sqrt{12}}$ ✔

$= \left[\dfrac{36}{2} - 2\sqrt{12} - \dfrac{4}{\sqrt{12}}\right] - \left[\dfrac{9}{2} - 2\sqrt{3} - \dfrac{4}{\sqrt{3}}\right]$

$= \left[18 - 2\sqrt{12} - \dfrac{\sqrt{12}}{3}\right] - \left[\dfrac{9}{2} - 2\sqrt{3} - \dfrac{4\sqrt{3}}{3}\right]$ ✔

$= 18 - 4\sqrt{3} - \dfrac{2\sqrt{3}}{3} - \dfrac{9}{2} + 2\sqrt{3} + \dfrac{4\sqrt{3}}{3}$

$= \dfrac{27}{2} - \dfrac{4\sqrt{3}}{3}$ ✔

(Total for Question 10 is 4 marks)

10

11 (a) Find the first 3 terms, in ascending powers of x, of the binomial expansion of

$(2 - 3x)^5$

giving each term in its simplest form.

(4)

(b) Explain how you would use your expansion to give an estimate for the value of 1.94^5.

(1)

(a) $(2 - 3x)^5 = (2)^5 + 5(2)^4(-3x) + 10(2)^3(-3x)^2 + ...$ ✓

$= 32 - 240x + 720x^2$ ✓✓✓ (each term)

(b) Let $2 - 3x = 1.94$

$x = 0.02$

then substitute into the expansion. ✓

Revision Guide
pages 24, 25

Hint Q11a

Be extra careful with negative terms. Always write every part of each term in brackets before you start to work it out.

Hint Q11b

$2 - 3x = 1.94$

(Total for Question 11 is 5 marks)

11

Revision Guide
page 7

Problem solving

For any question that says 'prove that' or 'show that', always show every step of your working. You should explain what you are doing at each stage and your last line of working should be the statement that was to be proved.

Hint

Use the discriminant.

12 Prove that the roots of the equation $x^2 - 2px + p^2 - q^2 = 0$ where p and q are constants, are real.

(3)

Discriminant $= b^2 - 4ac$ where $a = 1$, $b = -2p$, $c = p^2 - q^2$ ✔

$\qquad = (-2p)^2 - 4(p^2 - q^2)$

$\qquad = 4p^2 - 4p^2 + 4q^2$ ✔

$\qquad = 4q^2$ which must be positive so real roots ✔

(Total for Question 12 is 3 marks)

12

13

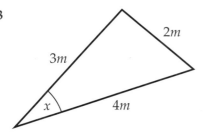

2m

3m

Not drawn to scale

x   4m

Revision Guide
page 27

A student was asked to find the value of cos x in terms of m from the triangle in the diagram.

The student's attempt is shown below:

$2m^2 = 4m^2 + 3m^2 - 2 \times 4m \times 3m \cos x$

$-5m^2 = -24m \cos x$

$\cos x = \dfrac{5m}{24}$

(a) Identify the errors made by the student.

(2)

(b) Find the correct value of $\cos x$.

(3)

(c) Use this value to find an exact value for $\sin x$.

(2)

Hint Q13a

Think about how you would work out the answer yourself, then compare this method with the sample answer given.

Hint Q13c

Draw a right-angled triangle with lengths corresponding to $\cos x$ on the adjacent and hypotenuse. Use Pythagoras to find the opposite and then use this to find $\sin x$. An exact value will include a surd for this question.

(a) Need brackets round $(2m)^2$, $(4m)^2$ and $(3m)^2$ so that these

   terms become $4m^2$, $16m^2$ and $9m^2$. ✔

   $2 \times 4m \times 3m = 24m^2$ not $24m$ ✔

(b) $a^2 = b^2 + c^2 - 2bc \cos A$ ✔

   $4m^2 = 16m^2 + 9m^2 - 2 \times 4m \times 3m \cos x$ ✔

   $-21m^2 = -24m^2 \cos x$

   $\cos x = \dfrac{7}{8}$ ✔

(c)  $\text{opp} = \sqrt{8^2 - 7^2}$ ✔

         $= \sqrt{15}$

      $\sin x = \dfrac{\sqrt{15}}{8}$ ✔

(Total for Question 13 is 7 marks)

13

Revision Guide
page 35

Problem solving

Use the formula for differentiation from first principles given in the formulae booklet. You need to state what happens to the expression as $h \to 0$, and write down the result you have proved.

14 Prove, from first principles, that the derivative of $\dfrac{1}{x}$ is $-\dfrac{1}{x^2}$.

(4)

$$\frac{f(x + h) - f(x)}{h} = \frac{\dfrac{1}{x + h} - \dfrac{1}{x}}{h} \quad \checkmark$$

$$= \frac{x - (x + h)}{hx(x + h)} \quad \checkmark$$

$$= \frac{-h}{hx(x + h)}$$

$$= \frac{-1}{x^2 + hx} \quad \checkmark$$

As $h \to 0$, $\dfrac{-1}{x^2 + hx} \to \dfrac{-1}{x^2}$, so the derivative of $\dfrac{1}{x}$ is $\dfrac{-1}{x^2}$ $\quad \checkmark$

(Total for Question 14 is 4 marks)

14

15 A hollow container in the shape of a cuboid has a base that measures $3x$ cm by x cm. The container doesn't have a top and has a height of y cm. The external surface area is 200 cm^2.

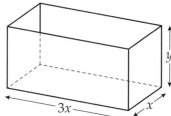

Revision Guide
page 41

(a) Show that the volume, $V \text{ cm}^3$, of the container is given by

$$V = 75x - \frac{9x^3}{8}$$

(4)

(b) Given that x can vary, find the maximum value of V to the nearest cm^3 and justify that this value is a maximum.

(7)

Modelling

For part (a), form an equation from the surface area information with y as the subject. Form an equation for V in terms of x and y and then substitute in for y.

Problem solving

In part (b), you will need to differentiate to find the maximum value. Don't forget to differentiate a second time to show that it's a maximum.

LEARN IT!

Justify means **give reasons**.

(a) Surface area $= 3x^2 + 2(3xy) + 2(xy)$

$= 3x^2 + 8xy = 200$ ✔

$y = \dfrac{200 - 3x^2}{8x}$ ✔

Volume $= 3x^2 y$

$= 3x^2 \times \dfrac{(200 - 3x^2)}{8x}$ ✔

$V = 75x - \dfrac{9x^3}{8}$ ✔

(b) $\dfrac{dV}{dx} = 75 - \dfrac{27x^2}{8}$ ✔

$0 = 75 - \dfrac{27x^2}{8}$ ✔

$\dfrac{27x^2}{8} = 75$

$x^2 = \dfrac{200}{9}$ ✔

$x = \dfrac{10\sqrt{2}}{3} = 4.71 \text{ cm}$ ✔

$\dfrac{d^2V}{dx^2} = -\dfrac{27x}{4}$ ✔

So the maximum value of V occurs when $x = 4.71 \text{ cm}$ (3 s.f.)

When $x = 4.71$, $\dfrac{d^2V}{dx^2} = -34.82$, so maximum ✔

Hence $V_{max} = 75 \times 4.71 - \dfrac{9 \times (4.71)^3}{8}$

$= 236 \text{ cm}^3$ ✔

(Total for Question 15 is 11 marks)

15

Revision Guide
page 52

Modelling

In part (a), you will need to use ln (natural logarithms) to find k.

Hint Q16c

Think about what happens as *t* tends to infinity.

LEARN IT!

$$\frac{d(e^{kt})}{dt} = ke^{kt}$$

This is a standard differential.

16 The total sales of a new novel in thousands of books, *B*, *t* months after release are modelled by the formula

$$B = 50(1 - e^{kt})$$

In the first month it is estimated that 10 000 books are sold.

(a) Find *k*.

(3)

(b) How many books do the publishers expect to sell in the first 5 months?

(2)

(c) Show that, according to this model, no more than 50 000 copies will be sold in the lifetime of the book.

(2)

(d) Show that the rate of sales of the book after *t* months is given by $\frac{dB}{dt} = Ae^{kt}$, where *A* is a constant to be found to 3 significant figures.

(2)

(a) $10 = 50(1 - e^k)$

$0.2 = 1 - e^k \Rightarrow e^k = 0.8$ ✓

$\ln e^k = \ln 0.8$ ✓

$k = \ln 0.8$ ✓

(b) $B = 50(1 - e^{5\ln 0.8})$ ✓

$= 33.616 \Rightarrow 33\,616$ books ✓

(c) As $t \to \infty$, $e^{t\ln 0.8} \to 0$ ✓

$\Rightarrow B \to 50 \Rightarrow$ Books = 50000 ✓

(d) $B = 50(1 - e^{-0.223t})$

$= 50 - 50e^{-0.223t}$

$\frac{dB}{dt} = -50 \times -0.223e^{-0.223t}$ ✓

$= 11.2\,e^{-0.223t}$ ✓

(Total for Question 16 is 9 marks)

TOTAL FOR PAPER IS 100 MARKS

16

Paper 2: Statistics and Mechanics
SECTION A: STATISTICS

Answer ALL questions. Write your answers in the spaces provided.

Revision Guide pages 71, 72

1   The Venn diagram shows the probabilities that a randomly chosen student at a college will take part in basketball and cricket.

B represents the event that a student takes part in basketball.

C represents the event that a student takes part in cricket.

x and y are probabilities.

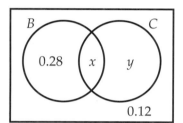

The events B and C are statistically independent.

(a) Write down the probability that a student plays cricket.

(1)

(b) Find the values of x and y.

(4)

Hint Q1a

$P(C) = 1 - P(\text{not } C)$

Hint Q1b

$P(B \text{ and } C) = P(B) \times P(C)$

Hint Q1b

Use the rule and solve for x.

(a) $1 - 0.28 - 0.12 = 0.6$ ✓

(b) $x = (0.28 + x) \times 0.6$ ✓

        $x = 0.168 + 0.6x$ ✓

    $0.4x = 0.168$

        $x = 0.42$ ✓

        $y = 0.6 - 0.42 = 0.18$ ✓

(Total for Question 1 is 5 marks)

17

Revision Guide
pages 56, 57, 60

LEARN IT!

In a systematic sample, elements are chosen at regular intervals.

Problem solving

You need to be familiar with the large data set, though you won't need to remember exact values from it.

Hint Q2c

Use the formulae for the mean and standard deviation and give your answers to 3 s.f.

LEARN IT!

Mean $= \dfrac{\sum x}{n}$

Standard deviation
$= \sqrt{\dfrac{\sum x^2}{n} - \left(\dfrac{\sum x}{n}\right)^2}$

2 Lucy is investigating the daily minimum temperature for Leeming in the months of May to August (inclusive) 2015.

She uses the large data set for her investigation.

(a) Describe how Lucy can take a systematic sample of 30 days.

(3)

(b) From your knowledge of the large data set, explain why this might not give Lucy a sample of size 30.

(1)

The data Lucy collected are summarised as follows:

$$n = 30 \qquad \sum x = 248 \qquad \sum x^2 = 2361$$

(c) Calculate the mean and standard deviation of Lucy's sample.

(3)

(a) $120 \div 30 = 4$ ✓

Select a random number from 1 to 4 and use it for the first

value to be chosen. ✓

Select every fourth value thereafter. ✓

(b) There could be gaps in the data, i.e. days when the minimum

temperature was not recorded. ✓

(c) Mean $= \dfrac{248}{30} = 8.27$ (3 s.f.) ✓

Standard deviation $= \sqrt{\dfrac{2361}{30} - 8.27^2} = 3.22$ (3 s.f.) ✓✓

(Total for Question 2 is 7 marks)

18

3   A company requires applicants for jobs to take an aptitude test. The possible outcomes are: pass, fail, be retested. The probabilities for these outcomes are:

P(pass) = 0.3   P(fail) = 0.5   P(be retested) = 0.2

An applicant can be retested twice. If they do not pass after being retested twice, they are rejected.

The probabilities for the outcomes with the first retesting are:

P(pass) = 0.3   P(fail) = 0.5   P(be retested) = 0.2

The probabilities for the outcomes with the second retesting are:

P(pass) = 0.4   P(fail) = 0.6

(a) Draw a tree diagram to illustrate these outcomes.

(3)

(b) Find the probability that a randomly selected applicant is accepted.

(2)

(c) Three friends apply for jobs with this company. What is the probability that two of them pass at the first attempt and one of them at the final attempt?

(3)

Revision Guide
page 73

Hint Q3a

This will be a 3-stage tree but the 2nd and 3rd stages only follow the 'retested' branches.

Hint Q3b

There are three ways an applicant can be successful. Work out each probability and add them together.

Hint Q3c

Be careful – consider how many ways this can be done.

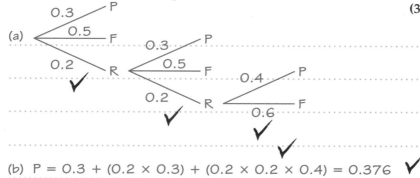

(b) P = 0.3 + (0.2 × 0.3) + (0.2 × 0.2 × 0.4) = 0.376 ✓

(c) The friend who passes at the final attempt can be chosen in

three different ways:

1   pass, pass, retest pass (PPR)

2   PRP

3   RPP ✓

The overall probability is

P = 3 × 0.3 × 0.3 × (0.2 × 0.2 × 0.4) = 0.00432 ✓

(Total for Question 3 is 8 marks)

Revision Guide
pages 75, 76, 77

Hint Q4a

You need to know when a binomial distribution can be used.

Problem solving

Use the binomial probability distribution to express the problem to be solved in terms of clearly defined mathematical statements. Finding the critical region enables you to comment on actual outcomes.

Hint Q4b

Write down the parameters you are going to use.

Hint Q4c

Be careful. This is a two-tailed test.

Hint Q4d

5% in each tail. $n = 20$, $p = 0.4$. Look for probabilities < 0.05

4 A coffee shop provides its customers with wi-fi.

The probability that a randomly selected customer uses this wi-fi is 0.4. Twelve customers are selected at random.

(a) Give two reasons why the number of customers in the sample who use wi-fi can be modelled using a binomial distribution.

(2)

(b) Find the probability that at least six of the customers in the sample used the wi-fi.

(2)

Another branch of this coffee shop also provides its customers with free wi-fi. The manager suspects that the probability of a customer using this facility may be different from 0.4. A random sample of 20 customers is selected.

(c) Write down the hypotheses that should be used to test the manager's theory.

(1)

(d) Using a 10% level of significance, find the critical region for a two-tailed test to investigate the manager's theory. You should state the probability of rejection in each tail, which should be less than 0.05.

(3)

(e) Find the actual significance level of the test based on your critical region from part (d).

(1)

One morning, the manager finds that 11 out of 20 customers are using the wi-fi.

(f) Comment on the manager's theory in light of this observation.

(1)

(a) Any two of: fixed number of trials (12), independent trials, two possible outcomes, fixed probability of 'success' (0.4) ✓✓

(b) Using B(12, 0.4), $P(X \geqslant 6)$ (or $1 - P(X \leqslant 5)$) = 0.3348 ✓✓

(c) $H_0: p = 0.4$ and $H_1: p \neq 0.4$ ✓

(d) Using B(20, 0.4), $P(X \leqslant 3) = 0.0160$ ✓

and $P(X \geqslant 13) = 1 - P(X \leqslant 12) = 0.0210$ ✓

$\Rightarrow$ Critical region is $x \leqslant 3$ and $x \geqslant 13$ with the probabilities

of rejection in each tail as above. ✓

20

(e) 0.0160 + 0.0210 = 0.0370 or 3.7% ✓

(f) 11 is not in the critical region so there is insufficient evidence

to support the manager's theory. ✓

Hint Q4e

Add the probabilities from your critical region.

Hint Q4f

Look at how the value 11 relates to the critical region.

Modelling

Always interpret your answers in the context of the question. Don't just leave them as expressions involving mathematical symbols.

(Total for Question 4 is 10 marks)

Revision Guide
page 81

Hint Q5a

Label your sketch with the given information. Time goes on the horizontal axis.

Hint Q5b

The area under a velocity–time graph represents distance.

SECTION B: MECHANICS

Answer ALL questions. Write your answers in the spaces provided.

Unless otherwise indicated, whenever a numerical value of g is required, take $g = 9.8\,\mathrm{m\,s^{-2}}$ and give your answer to either 2 significant figures or 3 significant figures.

5 A particle moves in a straight line. At time $t = 0$, the velocity is $u\,\mathrm{m\,s^{-1}}$. The particle then accelerates with constant acceleration to $11\,\mathrm{m\,s^{-1}}$ in $3\,\mathrm{s}$.

The particle maintains this velocity of $11\,\mathrm{m\,s^{-1}}$ for another $10\,\mathrm{s}$.

It then decelerates with constant deceleration to rest in a further $2\,\mathrm{s}$.

(a) Sketch a velocity–time graph to illustrate the motion of the particle.

(3)

(b) If the total distance travelled is $148\,\mathrm{m}$, work out the value of u.

(3)

(a)

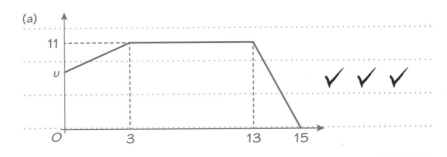

(b) Area of rectangle = 10 × 11 = 110

Area of triangle = $\dfrac{1}{2}$ × 2 × 11 = 11

Area of trapezium = 148 − 110 − 11 = 27 ✓

$\dfrac{3}{2}$ × (u + 11) = 27 ✓

u = 7 m s⁻¹ ✓

(Total for Question 5 is 6 marks)

22

6 A particle moves along the x-axis with velocity $v \, \text{m s}^{-1}$.

At time t seconds, the velocity of the particle is given by

$$v = 35t - 10t^2$$

The positive direction is in the sense of increasing x.

Find the distance travelled between the times when the particle is instantaneously at rest.

(5)

Instantaneously at rest when $35t - 10t^2 = 0$ ✓

$5t(7 - 2t) = 0$, giving $t = 0$ and $t = 3.5$ ✓

$\text{Distance} = \displaystyle\int_0^{3.5} 35t - 10t^2 \, dt$ ✓

$= \left[\dfrac{35t^2}{2} - \dfrac{10t^3}{3} \right]_0^{3.5}$ ✓

$= (214.375 - 142.9166...) - (0)$

$= 71.5 \, \text{m (3 s.f.)}$ ✓

Revision Guide
pages 91, 92

Problem solving

You are given an expression for v in terms of t but the question asks for distance travelled, with certain conditions applied, so you need to work out a strategy.

Hint

You can't find the distance (by integration) until you first work out the times when the particle is instantaneously at rest.

(Total for Question 6 is 5 marks)

Revision Guide
pages 82, 83

Hint Q7a

Use appropriate **suvat** formulae to set up two simultaneous equations in *a* and *u*.

Hint Q7b

You know *u*, *t* and *a*, and you need to find *s*.

7   A car is driven with constant acceleration, $a\,\text{m s}^{-2}$, along a straight road.

Its speed when it passes a road sign is $u\,\text{m s}^{-1}$.

In the first 3 seconds after passing the sign, the car travels $36\,\text{m}$.

5 seconds after passing the sign, the car has a speed of $26\,\text{m s}^{-1}$.

(a) Write down two equations connecting a and u. Hence find the values of a and u.

(5)

(b) Find the total distance travelled by the car in the first 5 seconds after it passes the sign.

(2)

(a) Using $s = ut + \frac{1}{2}at^2$, $s = 36$ when $t = 3$ gives

$36 = 3u + 4.5a$ ① ✓

Using $v = u + at$, $v = 26$ when $t = 5$ gives

$26 = u + 5a$ ② ✓

Solving simultaneously:

$78 = 3u + 15a$ ② × 3 ✓

$- \;\; 36 = 3u + 4.5a$ ①

$42 = \qquad 10.5a$

So $a = 4$ and, substituting, $u = 6$ ✓✓

(b) Using $s = ut + \frac{1}{2}at^2$ with $u = 6$, $a = 4$ and $t = 5$: ✓

$s = 6 \times 5 + \frac{1}{2} \times 4 \times 5^2$

$= 30 + 50 = 80\,\text{m}$ ✓

(Total for Question 7 is 7 marks)

24

8   A block *A* of mass 4 kg is held at rest on a rough horizontal table.

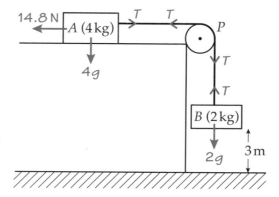

It is attached to one end of a light inextensible string.

The string passes over a small smooth pulley *P* fixed at the edge of the table.

Hint Q8a

Put in all the forces acting on the blocks, then use Newton's 3rd law for both particles and solve for *a*.

The other end of the string is attached to a block *B* of mass 2 kg, hanging freely below *P* and with *B* at a height of 3 m above the horizontal floor.

The system is released from rest with the string taut.

The resistance to the motion of *A* from the rough table is modelled as having constant magnitude 14.8 N.

A and *B* are modelled as particles.

(a) Show that the acceleration of *B* is 0.8 m s^{-2}.

(3)

(b) Calculate the tension in the string.

(2)

(c) Calculate the speed with which *B* hits the floor.

(2)

Block *A* never reaches the pulley at the edge of the table.

(d) After *B* has hit the floor, calculate the further distance *A* travels before coming to rest.

(4)

(e) State how, in your calculations, you have used the fact that the string is inextensible.

(1)

Problem solving

Label your diagram with all the forces acting on the two blocks. The pulley is smooth, so the tension in the string is the same on each side of the pulley. Once *B* has hit the floor you will need to re-visit the forces acting on *A*.

Modelling

Initially the blocks are connected by a light inextensible string and they are modelled as particles.

Hint Q8b

Substitute to find *T*.

(a) Using Newton's 3rd law for A: $T - 14.8 = 4a$ ✔

Using Newton's 3rd law for B: $2g - T = 2a$ ✔

Adding gives: $19.6 - 14.8 = 6a$

so $6a = 4.8 \Rightarrow a = 0.8\,\text{m s}^{-2}$ ✔

(b) Substituting $a = 0.8$: ✔ $T = 14.8 + 4(0.8) = 18\,\text{N}$ ✔

Hint Q8c

A **suvat** equation is needed here.

Aiming higher

In part (d), you must find the deceleration of A before you can use the appropriate **suvat** equation. Think about the forces acting on A after B hits the floor and the string becomes slack.

Hint Q8e

Think back to the original motion of the blocks.

(c) Using $v^2 = u^2 + 2as$ with $u = 0$, $a = 0.8$ and $s = 3$:

$$v^2 = 0 + 2 \times 0.8 \times 3 = 4.8 \checkmark$$

$$v = \sqrt{4.8} = 2.1908... = 2.19\,\text{m}\,\text{s}^{-1} \text{ (3 s.f.)} \checkmark$$

(d) A now moves independently of B and decelerates

Using Newton's 3rd law:

$-14.8 = 4a'$ where a' is the deceleration $\checkmark$

giving $a' = -3.7\,\text{m}\,\text{s}^{-2}$ $\checkmark$

Using $v^2 = u^2 + 2as$ with $u = 2.1908...$, $v = 0$

and $a = a' = -3.7$:

$0 = 4.8 - 7.4s$ $\checkmark$

giving $s = 0.6486... = 0.65\,\text{m}$ (2 s.f.) $\checkmark$

So A travels another 0.65 m before coming to rest.

(e) The blocks are initially connected by a light inextensible string, which implies that the acceleration of both blocks is the same. So if the acceleration of B is $0.8\,\text{m}\,\text{s}^{-2}$, then A also accelerates at $0.8\,\text{m}\,\text{s}^{-2}$. $\checkmark$

(Total for Question 8 is 12 marks)

TOTAL FOR PAPER IS 60 MARKS

26

Paper 1: Pure Mathematics

Revision Guide
page 1

Answer ALL questions. Write your answers in the spaces provided.

1 (a) Express 27^{4k+2} in the form 9^m, giving m in terms of k.

(3)

Hint Q1a

Start by writing 27 as a power of 3, then write 3 as a power of 9.

(b) Find the value of n in the equation

$$\frac{8^{2n+1}}{4^n} = 32$$

(3)

Hint Q1b

Change 8, 4 and 32 into powers of 2.

(a) $27^{4k+2} = (3^3)^{4k+2}$ ✓

$= 3^{12k+6}$

$= \left(9^{\frac{1}{2}}\right)^{12k+6}$ ✓

$= 9^{6k+3}$ ✓

(b) $\dfrac{8^{2n+1}}{4^n} = 32$

$\dfrac{2^{3(2n+1)}}{2^{2n}} = 2^5$

$2^{3(2n+1)-2n} = 2^5$ ✓

$3(2n+1) - 2n = 5$ ✓

$4n + 3 = 5$

$4n = 2$

$n = \dfrac{1}{2}$ ✓

(Total for Question 1 is 6 marks)

Revision Guide
page 4

Hint

Start by dividing the equation by a. Now complete the square and rearrange for x.

2 $ax^2 + bx + c = 0$.

Prove, by completing the square, that

$$x = \frac{-b \pm \sqrt{b^2 - 4ac}}{2a}$$

(6)

$$ax^2 + bx + c = 0$$

$$x^2 + \frac{bx}{a} + \frac{c}{a} = 0 \checkmark$$

$$\left(x + \frac{b}{2a}\right)^2 - \frac{b^2}{4a^2} + \frac{c}{a} = 0 \checkmark$$

$$\left(x + \frac{b}{2a}\right)^2 = \frac{b^2}{4a^2} - \frac{c}{a} \checkmark$$

$$x + \frac{b}{2a} = \sqrt{\frac{b^2}{4a^2} - \frac{4ac}{4a^2}} \checkmark$$

$$x = -\frac{b}{2a} \pm \frac{\sqrt{b^2 - 4ac}}{2a} \checkmark$$

$$x = \frac{-b \pm \sqrt{b^2 - 4ac}}{2a} \checkmark$$

(Total for Question 2 is 6 marks)

28

3 $f(x) = (2 + kx)^7$, where k is a constant.

Given that the coefficient of x^3 in the binomial expansion of $f(x)$ is 70, find the value of k.

(3)

Revision Guide
page 5

x^3 term $= 35\,(2^4)\,(kx)^3$ ✔

$35 \times 16 \times k^3 = 70$

$k^3 = \dfrac{1}{8}$ ✔

$k = \dfrac{1}{2}$ ✔

Hint

Use the **nCr** button on your calculator to find the coefficient of the term then multiply this by 2^4 and $(kx)^3$. This will give the x^3 term which can be equated to 70, then solve to find k.

(Total for Question 3 is 3 marks)

29

Revision Guide
pages 20, 22

Hint Q4a

Substitute $y = 2x + 11$ into the circle equation and show there is only one solution. If there is one repeated root, then the line must touch the circle at exactly one point, so it must be a tangent.

Hint Q4b

It might help to sketch the circle and the line. The first step will be to find the point of contact. Use the fact that the tangent to a circle is perpendicular to a radius at the point of contact.

4 (a) Show that the line $y = 2x + 11$ is a tangent to the circle

$$x^2 + y^2 - 6x - 4y = 32$$

(5)

(b) Find the equation of the diameter of the circle that passes through the point of contact of the tangent with the circle.

(3)

(a) $\quad x^2 + (2x + 11)^2 - 6x - 4(2x + 11) = 32 \checkmark$

$\quad x^2 + 4x^2 + 44x + 121 - 6x - 8x - 44 = 32 \checkmark$

$\quad\quad\quad\quad\quad\quad 5x^2 + 30x + 45 = 0 \checkmark$

$\quad\quad\quad\quad\quad\quad\quad x^2 + 6x + 9 = 0$

$\quad\quad\quad\quad\quad\quad\quad (x + 3)^2 = 0 \checkmark$

$\Rightarrow x = -3$ hence only one repeated solution so $y = 2x + 11$

must be a tangent to the circle. $\checkmark$

(b)

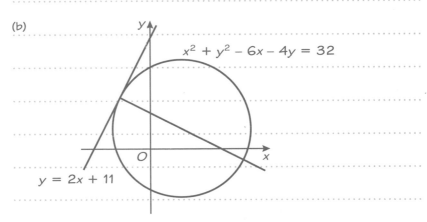

Point of contact is when $x = -3 \Rightarrow y = 2(-3) + 11 = 5$

Diameter is perpendicular to tangent so

$\text{gradient} = -\dfrac{1}{\text{gradient of the tangent}} = -\dfrac{1}{2} \checkmark$

$y - y_1 = m(x - x_1)$

$\Rightarrow y - 5 = -\dfrac{1}{2}(x - -3) \checkmark$

$\quad\quad y = -\dfrac{1}{2}x + \dfrac{7}{2} \checkmark$

(Total for Question 4 is 8 marks)

30

81

5   Prove that

$$\frac{1 - \sin\theta}{1 + \sin\theta} \equiv \left(\frac{1}{\cos\theta} - \tan\theta\right)^2$$

(5)

Revision Guide
page 31

$$\frac{1 - \sin\theta}{1 + \sin\theta} \equiv \frac{1 - \sin\theta}{1 + \sin\theta} \times \frac{1 - \sin\theta}{1 - \sin\theta}$$

$$\equiv \frac{(1 - \sin\theta)(1 - \sin\theta)}{(1 + \sin\theta)(1 - \sin\theta)} \checkmark$$

$$\equiv \frac{(1 - \sin\theta)^2}{1 - \sin^2\theta} \checkmark$$

$$\equiv \frac{(1 - \sin\theta)^2}{\cos^2\theta} \checkmark$$

$$\equiv \left(\frac{1 - \sin\theta}{\cos\theta}\right)^2 \checkmark$$

$$\equiv \left(\frac{1}{\cos\theta} - \frac{\sin\theta}{\cos\theta}\right)^2$$

$$\equiv \left(\frac{1}{\cos\theta} - \tan\theta\right)^2 \checkmark$$

LEARN IT!

You will need to use the trigonometrical identities $\sin^2\theta + \cos^2\theta \equiv 1$ and $\frac{\sin\theta}{\cos\theta} \equiv \tan\theta$.

Hint

Multiply the left-hand side by $\frac{1 - \sin\theta}{1 - \sin\theta}$
Then work through from the left-hand side to the right-hand side.

(Total for Question 5 is 5 marks)

31

Revision Guide
page 15

Hint Q6a

The reciprocal graph is a transformation of the graph $y = \dfrac{2}{x}$

Hint Q6b

Write $\dfrac{2}{x + 1} = 2x + 3$
and rearrange to form a quadratic. Telling you that you need to round to 2 d.p. is a giveaway that you need to use the quadratic formula.

6 (a) On the same axes sketch the graphs

$$y = \frac{2}{x + 1}$$

$$y = 2x + 3$$

Show where the graphs cross the axes and any asymptotes.

(4)

(b) Find the coordinates of the points of intersection of the two graphs. Round your answers to 2 d.p.

(3)

(a)

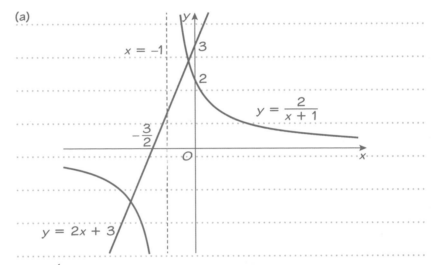

✔ for lines in the correct places

✔ for asymptote at $x = -1$

✔✔ for three points of intersection with axes

(b) $\dfrac{2}{x + 1} = 2x + 3$

$2 = (x + 1)(2x + 3)$ ✔

$2 = 2x^2 + 5x + 3$

$2x^2 + 5x + 1 = 0$ ✔

$x = \dfrac{-5 \pm \sqrt{5^2 - 8}}{4}$

$= -0.22 \text{ or } -2.28$

$\Rightarrow y = 2.56 \text{ or } -1.56$ ✔

So points of intersection are $(-0.22, 2.56)$ and $(-2.28, -1.56)$

(Total for Question 6 is 7 marks)

7 (a) Given that $f(x) = \dfrac{(2x + 3)(x - 5)}{x}$, $x \neq 0$, find $f'(x)$.

(3)

(b) Show that $f(x)$ is increasing for the interval $1 < x < 2$

(2)

(a) $f(x) = \dfrac{2x^2 - 7x - 15}{x}$

$= 2x - 7 - 15x^{-1}$ ✔

$f'(x) = 2 + 15x^{-2}$ ✔✔

$= 2 + \dfrac{15}{x^2}$

(b) $x^2 \geqslant 0$ for all real values of x, so $f(x) = 2 + \dfrac{15}{x^2}$ is positive for all values of x in the given interval.

Hence the function is increasing on this interval. ✔✔

Revision Guide
page 39

Hint Q7a

Expand the brackets and simplify the expression first.

LEARN IT!

A function $f(x)$ is increasing if $f'(x) \geqslant 0$ for all values of x in the given interval.

(Total for Question 7 is 5 marks)

33

Revision Guide
page 2

Hint

Remove a common factor then use the difference of two squares, done twice.

8   Factorise fully

$2x^5 - 32x$

(3)

$2x^5 - 32x = 2x(x^4 - 16)$ ✓

$= 2x(x^2 - 4)(x^2 + 4)$ ✓

$= 2x(x - 2)(x + 2)(x^2 + 4)$ ✓

(Total for Question 8 is 3 marks)

34

9 Figures 1 and 2 show the curve with equation $y = 3x^2 + 2x - 1$

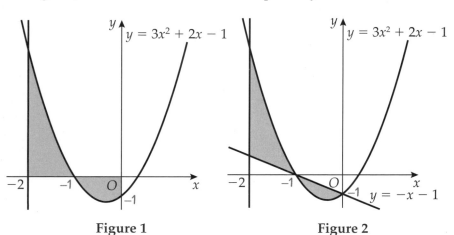

Figure 1   **Figure 2**

Revision Guide
pages 45, 46

Hint Q9a

Factorise to find the points of intersection with the x-axis, then sketch the quadratic graph.

Problem solving

For part (b), add $x + y = -1$ to your sketch graph and work out points of intersection so you can see the area required.

(a) Find the total shaded area shown in Figure 1, enclosed by the curve, the x-axis and the lines $x = -2$ and $x = 0$.

(4)

(b) Hence or otherwise find the shaded area shown in Figure 2 enclosed by the curve, the line $x + y = -1$ and the lines $x = -2$ and $x = 0$.

(3)

(a) $y = (3x - 1)(x + 1)$ so when $y = 0$, $x = \dfrac{1}{3}$ or $x = -1$

Total area $= \displaystyle\int_{-2}^{-1} (3x^2 + 2x - 1)dx + \int_{-1}^{0} (3x^2 + 2x - 1)dx$ ✔

Area 1 $= \displaystyle\int_{-2}^{-1} (3x^2 + 2x - 1)dx$

$= [x^3 + x^2 - x]_{-2}^{-1}$

$= (-1 + 1 + 1) - (-8 + 4 + 2) = 3$ ✔

Area 2 $= \displaystyle\int_{-1}^{0} (3x^2 + 2x - 1)dx$

$= [x^3 + x^2 - x]_{-1}^{0}$

$= (0) - (-1 + 1 + 1) = -1$ ✔

Total shaded area $= 3 + 1 = 4$ square units ✔

(b) When $x = -2$, $y = 1$ on the line $y = -x - 1$ ✔

Area of triangle above x-axis $= \dfrac{1}{2} \times 1 \times 1 = \dfrac{1}{2}$ ✔

Area of triangle below x-axis $= \dfrac{1}{2} \times 1 \times 1 = \dfrac{1}{2}$

New shaded area $= 4 - 1 = 3$ ✔

(Total for Question 9 is 7 marks)

35

Revision Guide
page 26

Problem solving

You can write a general multiple of 4 as 4k, where k is a positive integer. Find a general expression for the sum of three consecutive multiples of 4 and show that it has 12 as a factor.

10 Prove that the sum of three consecutive multiples of 4 is always a multiple of 12.

(3)

Three consecutive multiples of 4 can be written as 4k, 4(k + 1)

and 4(k + 2), where k is a positive integer. ✓

So their sum is given by

4k + 4(k + 1) + 4(k + 2) = 4k + 4k + 4 + 4k + 8

= 12k + 12

= 12(k + 1) ✓

Since k + 1 is a positive integer, this is a multiple of 12,

so the sum of three consecutive multiples of 4 must be a

multiple of 12. ✓

(Total for Question 10 is 3 marks)

36

87

11 (a) Find the equation of the straight line that passes through the points A $(-2, 7)$ and B $(3, 5)$, in the form $ax + by + c = 0$.

(3)

(b) Find the equation of the perpendicular bisector of AB.

(4)

(c) Point C lies on the perpendicular bisector of AB and is vertically above point B. Find the area of the triangle ABC.

(5)

Revision Guide
pages 17, 18, 19

(a) Gradient $= m = \dfrac{y_2 - y_1}{x_2 - x_1} = \dfrac{5 - 7}{3 - (-2)} = \dfrac{-2}{5}$ ✓

$y - y_1 = m(x - x_1) \Rightarrow y - 7 = \dfrac{-2}{5}(x - -2)$ ✓

$2x + 5y - 31 = 0$ ✓

(b) Perpendicular line has gradient $m = -\dfrac{1}{-\frac{2}{5}} = \dfrac{5}{2}$ ✓

Midpoint of $AB = \left(\dfrac{-2 + 3}{2}, \dfrac{7 + 5}{2}\right) = \left(\dfrac{1}{2}, 6\right)$ ✓

$y - y_1 = m(x - x_1) \Rightarrow y - 6 = \dfrac{5}{2}\left(x - \dfrac{1}{2}\right)$ ✓

$10x - 4y + 19 = 0$ ✓

(c) C has x coordinate of 3. Substitute this into

$10x - 4y + 19 = 0$ ✓

$30 - 4y + 19 = 0$

$4y = 49 \Rightarrow y = 12\dfrac{1}{4}$ ✓

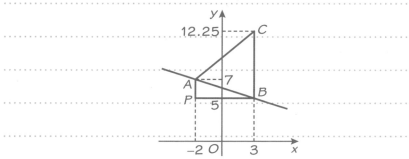

Area of trapezium $APBC = \dfrac{1}{2} \times 5 \times (2 + 7.25) = 23.125$ ✓

Area of triangle $APB = \dfrac{1}{2} \times 5 \times 2 = 5$ ✓

Area of triangle $ABC = 23.125 - 5 = 18.125$ ✓

(Total for Question 11 is 12 marks)

Hint Q11a

Use gradient $= \dfrac{y_2 - y_1}{x_2 - x_1}$
and then $y - y_1 = m(x - x_1)$
to find the equation of
the straight line, then
rearrange into the form
requested.

Hint Q11b

You will need the
perpendicular gradient
and the coordinates of
the midpoint of AB.

Problem solving

In part (c), point C
must have the same
x coordinate as B.
Substitute this value
in the equation of the
perpendicular to find
the coordinates of C.
Use the point $P(-2, 5)$
to make a trapezium
$APBC$ and a triangle
APB. Find the area
of the trapezium and
subtract the area of
the triangle.

37

Revision Guide
pages 42, 43

Hint Q12a

After integrating, substitute the given values to find the constant of integration.

Hint Q12b

Substitute $x = -1$ into f(x) using the constant of integration found in part (a).

12 $y = \int \left(2x^3 - 4 + \dfrac{1}{x^2}\right)dx$

 (a) Given that $y = 0$ when $x = 2$, find y as a function of x.

                 (3)

 (b) The point $P\,(-1, t)$ lies on the curve with equation $y = $ f(x). Find the value of t.

                 (2)

(a) $y = \dfrac{x^4}{2} - 4x - \dfrac{1}{x} + c$ ✓✓

  $0 = \dfrac{2^4}{2} - 4(2) - \dfrac{1}{2} + c$

  $c = \dfrac{1}{2}$ ✓

  $y = \dfrac{x^4}{2} - 4x - \dfrac{1}{x} + \dfrac{1}{2}$

(b) $t = \dfrac{(-1)^4}{2} - 4(-1) - \dfrac{1}{-1} + \dfrac{1}{2}$ ✓

  $t = \dfrac{1}{2} + 4 + 1 + \dfrac{1}{2} = 6$ ✓

            (Total for Question 12 is 5 marks)

38

13 The value of Andy's car can be modelled by the formula

$$V = 18\,600e^{-0.23t} + k$$

where £V is the value of the car, t is the age in years and k is a positive constant.

(a) The value of the car when new was £19 550. Find the value of k.

(1)

(b) Find the value of the car after 3 years, to the nearest £.

(2)

(c) Find the rate of decrease in the value of the car in £ per year to the nearest £ at the instant when the car is 5 years old.

(3)

(d) Sketch the graph of V against t.

(2)

(e) Interpret the meaning of the value of k.

(1)

(a) When $t = 0$, $19\,550 = 18\,600 + k \Rightarrow k = 950$ ✓

(b) $V = 18\,600e^{-0.69} + 950$ ✓

$V = £10\,279$ ✓

(c) $\dfrac{dV}{dt} = -0.23 \times 18\,600e^{-0.23t}$ ✓

$= -4278e^{-0.23t}$

When $t = 5$, $\dfrac{dV}{dt} = -4278e^{-1.15}$ ✓ $= £1355$ per year ✓

(d)

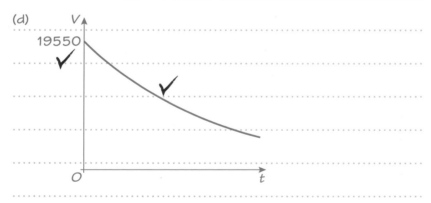

(e) As t increases $e^{-0.23t}$ tends to 0 so V tends to £950 which

is likely to be the scrappage value. ✓

(Total for Question 13 is 9 marks)

Revision Guide
page 52

Hint Q13a

$t = 0$ when the car is new.

Hint Q13b

After 3 years, $t = 3$. Use the value of k found in part (a).

Hint Q13c

Differentiate with respect to t.

Hint Q13d

Plot V on the vertical axis and t on the horizontal. Label V when $t = 0$ on the graph.

Modelling

In part (e), think about what happens to V as t increases.

Revision Guide
pages 48, 50

Hint Q14a

Take logs of both sides. Don't round your answers until the very end.

LEARN IT!

$\log(a^b) = b\log a$

$\log(a) - \log(b) = \log\dfrac{a}{b}$

14 (a) Solve $5^{2x} = 8^{x-1}$ giving your answer to 3 significant figures.

(3)

(b) Solve $\log_3(x + 1) - \log_3(x) = 2$.

(3)

(a) $\ln 5^{2x} = \ln 8^{x-1}$

$2x\ln 5 = (x - 1)\ln 8$ ✔

$2x\ln 5 - x\ln 8 = -\ln 8$

$x(2\ln 5 - \ln 8) = -\ln 8$ ✔

$x = \dfrac{-\ln 8}{2\ln 5 - \ln 8}$

$= -1.82$ ✔

Alternative solution

You can carry out the above process using logs to any base.

(b) $\log_3(x + 1) - \log_3(x) = 2$

$\log_3\left(\dfrac{x + 1}{x}\right) = 2$ ✔

$\left(\dfrac{x + 1}{x}\right) = 3^2$ ✔

$x + 1 = 9x$

$8x = 1$

$x = \dfrac{1}{8}$ ✔

(Total for Question 14 is 6 marks)

40

15 A parallelogram $ABCD$ has $\overrightarrow{AB} = \begin{pmatrix} 2 \\ 5 \end{pmatrix}$ and $\overrightarrow{AD} = \begin{pmatrix} 4 \\ 1 \end{pmatrix}$.

(a) Find $\left| \overrightarrow{BD} \right|$.

(2)

(b) Find angle $B\hat{A}D$.

(3)

(c) Find the area of parallelogram $ABCD$.

(2)

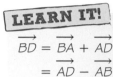

Revision Guide
pages 27, 33, 34

LEARN IT!

$\overrightarrow{BD} = \overrightarrow{BA} + \overrightarrow{AD}$
$= \overrightarrow{AD} - \overrightarrow{AB}$

(a)

$$\overrightarrow{BD} = \overrightarrow{AD} - \overrightarrow{AB} = \begin{pmatrix} 2 \\ -4 \end{pmatrix} \checkmark$$

$$\left| \overrightarrow{BD} \right| = \sqrt{2^2 + 4^2} = \sqrt{20} = 2\sqrt{5} = 4.47 \checkmark$$

(b) $\left| \overrightarrow{AB} \right| = \sqrt{2^2 + 5^2} = \sqrt{29}$ and $\left| \overrightarrow{AD} \right| = \sqrt{4^2 + 1^2} = \sqrt{17}$ $\checkmark$

$$\cos B\hat{A}D = \frac{29 + 17 - 20}{2 \times \sqrt{29} \times \sqrt{17}} = 0.585... \checkmark$$

$$B\hat{A}D = 54.2° \checkmark$$

(c) Area of a parallelogram is

$$|a||b|\sin C = \sqrt{29} \times \sqrt{17} \times \sin 54.2° \checkmark$$

Area = 18 square units $\checkmark$

Hint Q15b

There are other ways of doing this but finding the lengths of the three sides and using the cosine rule is the most straightforward.

LEARN IT!

Area of a triangle is $\frac{1}{2}|a||b|\sin C$ so area of a parallelogram is $|a||b|\sin C$.

(Total for Question 15 is 7 marks)

Revision Guide
pages 2, 37, 38

Problem solving

The wording makes this question look harder than it is. Sometimes you will need to pick out key words. You have seen the word 'tangent' so differentiate $y = ax^3 + x$ and then substitute $x = 1$. Then use the fact that you know the gradient of the tangent from its equation.

Hint Q16b

Use the value of a you found in part (a).

Aiming higher

In part (c), equate the two functions of x and rearrange to form a cubic equation. Use the knowledge that where the tangent touches the curve there are two solutions at the same place (so you know two of the solutions already). This will help you factorise the cubic expression.

16 The line L, defined by $y = 7x + k$, is a tangent to the curve C, defined by $y = ax^3 + x$, at the point where $x = 1$. a and k are both constants.

(a) Find the value of a.

(2)

(b) Find the value of k.

(2)

(c) Find the coordinates of the point where $y = 7x + k$ crosses the curve $y = ax^3 + x$.

(4)

(a) $\dfrac{dy}{dx} = 3ax^2 + 1$ ✔

When $x = 1$, $\dfrac{dy}{dx} = 3a + 1 = 7 \Rightarrow a = 2$ ✔

(b) $y = 2x^3 + x$ so when $x = 1$, $y = 3$ ✔

$y = 7x + k \Rightarrow 3 = 7 + k \Rightarrow k = -4$ ✔

(c) $2x^3 + x = 7x - 4$ ✔

$2x^3 - 6x + 4 = 0$

$\Rightarrow x^3 - 3x + 2 = 0$ ✔

$(x - 1)^2(x + 2) = 0$ ✔

When $x = -2$, $y = 7(-2) - 4 = -18$

L and C intersect when $x = -2$ so coordinates are $(-2, -18)$ ✔

(Total for Question 16 is 8 marks)

TOTAL FOR PAPER IS 100 MARKS

Paper 2: Statistics and Mechanics
SECTION A: STATISTICS

Revision Guide page 74

Answer ALL questions. Write your answers in the spaces provided.

1   The discrete random variable Y can take only values 1, 2, 3, 4 and 5.

Y has probability function

$$P(Y = y) = \begin{cases} k(4 - y)^2, & y = 1, 2, 3 \\ k(3y - 8), & y = 4, 5 \end{cases}$$

where k is a constant.

(a) Find the value of k and construct a table giving the probability distribution of Y.

(4)

(b) Find $P(Y \geqslant 4)$

(1)

Hint Q1a

Use the probability function for values of Y from 1 to 5. Remember that the sum of all probabilities = 1.

Hint Q1b

Use the appropriate values from your probability distribution.

(a) $P(Y = 1) = k(4 - 1)^2 = 9k$

$P(Y = 2) = 4k$

$P(Y = 3) = k$ ✔

$P(Y = 4) = k(3 \times 4 - 8) = 4k$

$P(Y = 5) = 7k$ ✔

Total probability $= 9k + 4k + k + 4k + 7k = 25k$,

so $25k = 1$, hence $k = \dfrac{1}{25}$ ✔

y	1	2	3	4	5
$P(Y = y)$	$\dfrac{9}{25}$	$\dfrac{4}{25}$	$\dfrac{1}{25}$	$\dfrac{4}{25}$	$\dfrac{7}{25}$

✔

(b) $P(Y \geqslant 4) = P(Y = 4) + P(Y = 5)$

$= \dfrac{4}{25} + \dfrac{7}{25} = \dfrac{11}{25}$ ✔

(Total for Question 1 is 5 marks)

Revision Guide
pages 58, 63

Hint Q2a

$n = 13$ so, for example, $\frac{n}{4} = 3.25$.

This is rounded up to 4, so Q_1 is the 4th observation from an ordered list.

Hint Q2b

Identify any outliers, deduce the whiskers, then draw the box plot using your Q_1, Q_2 and Q_3 values.

Aiming higher

The whiskers can be drawn either at the outlier boundaries, $Q_1 - 1.5 \times$ IQR and $Q_3 + 1.5 \times$ IQR, or at the smallest and largest data points which are not outliers.

2 Thirteen students sat a test, marked out of 50. These are their scores:

$$33, 25, 36, 40, 14, 49, 31, 17, 31, 44, 35, 28, 34$$

(a) Write down the values of Q_1, Q_2 and Q_3 for this data.

(3)

A value that is more than 1.5 times the interquartile range (IQR) above Q_3 or more than 1.5 times the IQR below Q_1 is called an outlier.

(b) Draw a box plot for this data.

(4)

(a) Rearrange the data in ascending order as

$$14, 17, 25, 28, 31, 31, 33, 34, 35, 36, 40, 44, 49$$

$n = 13$

$\frac{n}{4} = 3.25$, round up to 4, Q_1 is the 4th item, so $Q_1 = 28$ ✓

$\frac{n}{2} = 6.5$, round up to 7, Q_2 is the 7th item, so $Q_2 = 33$ ✓

$\frac{3n}{4} = 9.75$, round up to 10, Q_3 is the 10th item, so $Q_3 = 36$ ✓

(b) IQR $= 36 - 28 = 8$, $1.5 \times$ IQR $= 1.5 \times 8 = 12$

$Q_1 - 12 = 28 - 12 = 16$, so 14 is an outlier

$Q_3 + 12 = 36 + 12 = 48$, so 49 is an outlier ✓

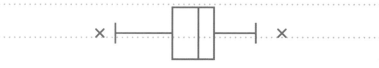

✓ for the two outliers

✓ for the whiskers drawn to 17 and 44

✓ for the box showing correct Q_1, Q_2 and Q_3

(Total for Question 2 is 7 marks)

3 A farmer collects data on annual rainfall, r cm, and the annual yield of broccoli, b tonnes per acre.

The table shows the results for 10 consecutive years.

r	75	79	78	84	74	76	80	85	77	81
b	4.5	4.8	4.7	5.1	4.6	4.6	4.9	5.2	4.7	4.8

The equation of the regression line of b on r is $b = 1.6 + 0.04r$.

(a) Give an interpretation of the value 0.04 in the regression equation.

(1)

(b) Comment on the reliability of using this regression equation to estimate the yield of broccoli in a year when the annual rainfall is

(i) 70 cm   (ii) 82 cm

(2)

(c) Explain why the regression equation is not suitable to estimate the annual rainfall in a year when the yield of broccoli is 4.5 tonnes per acre.

(1)

Revision Guide
pages 62, 68, 69

Hint Q3a

Relate your interpretation to the data, so write about amount of rainfall with respect to yield per acre.

Hint Q3b

Look at the range of the original data.

LEARN IT!

The response in part (c) here is a standard one. Consider independent and dependent variables.

(a) 0.04 represents the gradient of the regression line of b on r,

so for every increase of 1 cm in the annual rainfall the yield of

broccoli increases by 0.04 tonnes per acre. ✔

(b) (i) 70 cm is outside the range of the original data, so this is

an example of extrapolation which means that the estimate

will be unreliable. ✔

(ii) 82 cm is inside the range of the original data, so this is

an example of interpolation which means that the estimate

will be reliable. ✔

(c) The regression equation can only be used to work out the

dependent variable (b) given the value of the independent

variable (r), not the other way round. ✔

(Total for Question 3 is 4 marks)

45

Revision Guide
page 59

Hint

Times are given to the nearest minute, so 5–9 really means 4.5–9.5

Hint

Work out in which class intervals the 10th and 90th percentiles lie, then work out the positions within those intervals using linear interpolation.

Hint

Watch out! The class boundaries are not whole numbers.

4 In a survey, 120 shoppers in a supermarket were asked how many minutes, to the nearest minute, they had been in the store. The results are summarised in the table.

Number of minutes	Number of shoppers
1 – 4	5
5 – 9	20
10 – 19	28
20 – 29	51
30 – 59	16
Total	**120**

Use linear interpolation to estimate the 10% to 90% interpercentile range of this data.

(5)

10% of 120 = 12, so the 10th percentile is the 12th item and the 90th percentile is the 108th item.

The 12th item is in the 4.5 to 9.5 class interval: it is the 7th item in a class of width 5 ✓

$$10\text{th percentile} = 4.5 + \frac{7}{20} \times 5$$

$$= 4.5 + 1.75$$

$$= 6.25 \text{ minutes} ✓$$

The 108th item is in the 29.5 to 59.5 class interval: it is the 4th item in a class of width 30 ✓

$$90\text{th percentile} = 29.5 + \frac{4}{16} \times 30$$

$$= 29.5 + 7.5$$

$$= 37 \text{ minutes} ✓$$

10th to 90th interpercentile range = 37 − 6.25

$$= 30.75 \text{ minutes} ✓$$

(Total for Question 4 is 5 marks)

46

5 All cars more than three years old have to undergo an annual MOT test. 20% of cars fail the MOT test because of problems with their lights. This is the most common cause of failure nationally. A group of 5 randomly selected cars undergo an MOT test.

Revision Guide
pages 75, 76

(a) Give a reason why the binomial distribution is suitable for modelling the number of cars failing the MOT test due to a problem with their lights.

(1)

LEARN IT!

You need to know when a binomial distribution can be used.

(b) Find the probability that exactly 2 of them fail the lights test.

(2)

A car hire firm has 15 cars taking their MOT test. Six of them fail the lights test.

Hint Q5b

You need to use the number of ways of selecting 2 outcomes from 5 trials, with $p = 0.2$

(c) Using a 5% level of significance, find whether there is evidence to support the suggestion that cars from this firm fail the lights test more than the national average. You should state clearly the hypotheses that should be used.

(6)

Hint Q5c

When carrying out any hypothesis test, you should always write down your hypotheses clearly. This is a one-tailed test so your alternative hypothesis will be in the form

$H_1: p > \ldots$ or $H_1: p < \ldots$

(a) There are only two possible outcomes, pass or fail, and the probability of failing is fixed at 0.2 ✔

(b) P(exactly 2 fail) $= {}^5C_2 \times (0.2)^2 \times (0.8)^3$ ✔

$= 0.2048$ ✔

(c) $X \sim B(15, p)$, $H_0: p = 0.2$, $H_1: p > 0.2$

(this is a one-tailed test) ✔✔

Assume H_0 so that $p = 0.2$, and $X \sim B(15, 0.2)$.

$P(X \geqslant 6) = 1 - P(X \leqslant 5) = 0.0611$ ✔✔

Since $0.0611 > 0.05$ we do not reject the null hypothesis. ✔

There is insufficient evidence to support the suggestion that cars from this car hire firm fail the lights test more than the national average. ✔

Modelling

Always interpret your answers in the context of the question.

(Total for Question 5 is 9 marks)

47

Revision Guide
page 84

Hint Q6a

For vertical motion under gravity, always choose a positive direction (up or down). In this case, if you choose upwards as positive, then $a = -9.8\,\text{m}\,\text{s}^{-2}$ since gravity acts downwards.

Problem solving

If the stone starts at a height of h, then it will hit the ground when $s = -h$.

LEARN IT!

Standard modelling assumptions apply here.
$s = ut + \dfrac{1}{2}at^2$

Modelling

In part (b), you are asked for modelling assumptions. Think about the standard conditions that apply in questions involving vertical motion under gravity.

SECTION B: MECHANICS

Answer ALL questions. Write your answers in the spaces provided.

Unless otherwise indicated, whenever a numerical value of g is required, take $g = 9.8\,\text{m}\,\text{s}^{-2}$ and give your answer to either 2 significant figures or 3 significant figures.

6 A stone is thrown vertically upwards with speed $15\,\text{m}\,\text{s}^{-1}$ from a point h metres above the ground.

The stone hits the ground 4 seconds later.

(a) Find the value of h.

(3)

(b) State two modelling assumptions made when calculating your answer.

(2)

(a)

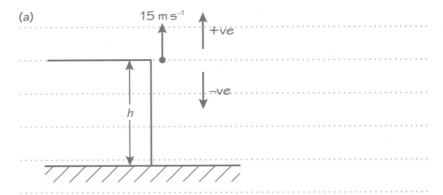

Taking upwards as the positive direction.

Using $s = ut + \dfrac{1}{2}at^2$, $s = -h$, $u = 15$, $t = 4$ and $a = -9.8$ ✓

$-h = 15 \times 4 - \dfrac{1}{2} \times 9.8 \times 4^2$ ✓

$-h = -18.4$, so $h = 18.4$ and the stone is thrown vertically upwards from a height of $18.4\,\text{m}$ ✓

(b) The stone is modelled as a particle. ✓

Gravity is constant and acts vertically downwards. ✓

(Total for Question 6 is 5 marks)

48

7   A lift of mass 220 kg is being lowered into a shaft by a vertical cable attached to the top of the lift.

The cable exerts an upward force of 2150 N on the lift.

A container of mass m kg rests on the floor of the lift.

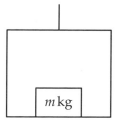

There is a constant upward resistance of 160 N on the lift and the normal reaction between the container and the lift is of magnitude 480 N.

The lift descends with constant acceleration, a m s^{-2}.

(a) Find the acceleration of the lift.

(4)

(b) Find the mass of the container.

(3)

(a) Applying Newton's 3rd law to the lift:

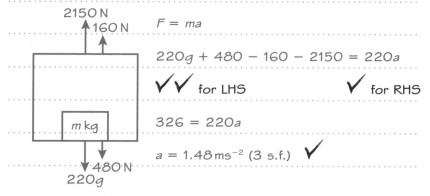

$F = ma$

$220g + 480 - 160 - 2150 = 220a$

✔✔ for LHS   ✔ for RHS

$326 = 220a$

$a = 1.48\,\text{ms}^{-2}$ (3 s.f.) ✔

(b) Applying Newton's 3rd law to the container:

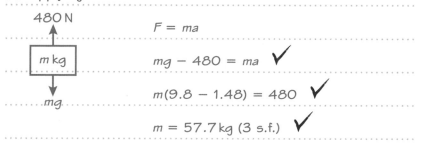

$F = ma$

$mg - 480 = ma$ ✔

$m(9.8 - 1.48) = 480$ ✔

$m = 57.7\,\text{kg}$ (3 s.f.) ✔

Alternative solution

You could use Newton's 3rd law on the whole system:

$220g + mg - 160 - 2150 = (220 + m)a$ ✔✔

giving $m = 57.7\,\text{kg}$, as above ✔

(Total for Question 7 is 7 marks)

Revision Guide
page 85

Hint Q7a

Use Newton's 3rd law. You can apply it to the lift, the container or the whole system. Choose whichever is the most appropriate.

Problem solving

Draw a diagram showing all the forces acting.

Hint Q7a

Put in all the forces acting on whatever part of the system you choose. Take care with signs – take downwards as positive.

Modelling

The system is such that all parts are moving in the same straight line, so you can treat the whole system as a single particle or consider parts of the system separately. The container and the lift remain in contact, so they exert equal and opposite forces on each other.

Hint Q7b

Another application of Newton's 3rd law is needed here.

Hint Q8a

The particle is in equilibrium so the resultant of the three forces must be the zero vector. Set up two simultaneous equations in a and b.

Problem solving

Forces are vectors, so in part (a) use the equilibrium condition to work out the values of a and b.

Hint Q8b

Find the resultant of $\mathbf{F}_1$ and $\mathbf{F}_3$. Then find its magnitude and direction. Use $F = ma$ to find the acceleration of the particle.

Problem solving

Because the particle was in equilibrium, it is accelerating from rest, so its acceleration will act in the same direction as the resultant force.

8 Three forces, $\mathbf{F}_1$, $\mathbf{F}_2$ and $\mathbf{F}_3$ act on a particle of mass 5 kg. The forces are given as:

$$\mathbf{F}_1 = 2a\mathbf{i} - 5b\mathbf{j}$$

$$\mathbf{F}_2 = -4b\mathbf{i} + 3a\mathbf{j}$$

$$\mathbf{F}_3 = 2\mathbf{i} - 2\mathbf{j}$$

where a and b are constants.

The particle is in equilibrium.

(a) Work out the values of a and b.

(3)

The force $\mathbf{F}_2$ is removed.

(b) Find the magnitude and bearing of the resulting acceleration of the particle.

(6)

(a) $\underset{\sim}{\mathbf{F}_1} + \underset{\sim}{\mathbf{F}_2} + \underset{\sim}{\mathbf{F}_3} = 0$ since the particle is in equilibrium.

$2a\underset{\sim}{\mathbf{i}} - 5b\underset{\sim}{\mathbf{j}} + (-4b\underset{\sim}{\mathbf{i}} + 3a\underset{\sim}{\mathbf{j}}) + 2\underset{\sim}{\mathbf{i}} - 2\underset{\sim}{\mathbf{j}} = 0$ ✓

Equating i coefficients   Equating j coefficients

$2a - 4b + 2 = 0$ $-5b + 3a - 2 = 0$ ✓

$2a - 4b = -2$ ① $3a - 5b = 2$ ②

$6a - 12b = -6$ ① × 3 $6a - 10b = 4$ ② × 2

Subtracting gives $2b = 10$, so $b = 5$

and substituting gives $a = 9$ ✓

(b) When $\underset{\sim}{\mathbf{F}_2}$ is removed, the resultant of $\underset{\sim}{\mathbf{F}_1}$ and $\underset{\sim}{\mathbf{F}_3}$ is

$18\underset{\sim}{\mathbf{i}} - 25\underset{\sim}{\mathbf{j}} + 2\underset{\sim}{\mathbf{i}} - 2\underset{\sim}{\mathbf{j}} = 20\underset{\sim}{\mathbf{i}} - 27\underset{\sim}{\mathbf{j}}$ ✓

$F = \sqrt{20^2 + 27^2} = 33.6\,\text{N}$ ✓

$33.6 = 5a$ ✓

$a = 6.72\,\text{ms}^{-2}$ ✓

$\theta = \tan^{-1}\dfrac{27}{20} = 53.47...°$ ✓

Bearing $= 90° + 53.5° = 143.5°$ ✓

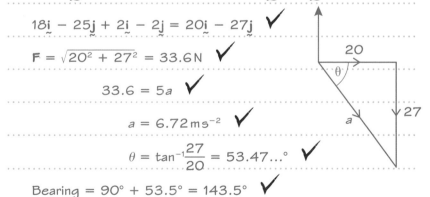

(Total for Question 8 is 9 marks)

9 A rabbit leaves its burrow at time $t = 0$ and runs alongside a straight fence before returning to its burrow.

Revision Guide
pages 91, 92

The rabbit is modelled as a particle moving in a straight line.

The distance, s metres, of the rabbit from its burrow at time t seconds is given by

$$s = \frac{1}{5}(t^4 - 16t^3 + 64t^2)$$

where $0 \leqslant t \leqslant 8$

(a) Explain the restriction $0 \leqslant t \leqslant 8$

(3)

(b) Show that the rabbit is initially at rest and find its distance from the burrow when it next comes to instantaneous rest.

(6)

Hint Q9a

Factorise and interpret the equation in the context of the problem.

Modelling

The rabbit is modelled as a particle moving in a straight line with its distance from its burrow, s, given as a function of time, t.

Hint Q9b

Instantaneous rest means $v = 0$. Use calculus to find the relevant value of t then use it to calculate the distance.

Problem solving

You will need calculus to connect expressions for s and v and consider times when the rabbit is instantaneously at rest.

(a) $s = \frac{1}{5}(t^4 - 16t^3 + 64t^2)$

$= \frac{1}{5}t^2(t^2 - 16t + 64)$

$= \frac{1}{5}t^2(t - 8)^2$ ✓

$s = 0$ when $t = 0$ and $t = 8$ ✓

For values of t in the range 0 to 8 the value of s is always

positive since t^2 and $(t - 8)^2$ are both > 0 ✓

(b) $v = \frac{ds}{dt}$ ✓

$v = \frac{1}{5}(4t^3 - 48t^2 + 128t)$

$= \frac{4}{5}t(t^2 - 12t + 32)$

$= \frac{4}{5}t(t - 4)(t - 8)$ ✓✓

$v = 0$ when $t = 0$, $t = 4$ and $t = 8$

So the rabbit is at rest when $t = 0$ and is next

instantaneously at rest when $t = 4$ ✓

$s = \frac{1}{5} \times 4^2 \times (4 - 8)^2$ ✓

$s = 51.2\,\text{m}$ ✓

(Total for Question 9 is 9 marks)

TOTAL FOR PAPER IS 60 MARKS

51

Formulae list

Pure Mathematics

Mensuration

Surface area of sphere $= 4\pi r^2$

Area of curved surface of cone $= \pi r \times$ slant height

Binomial series

$$(a + b)^n = a^b + \binom{n}{1}a^{n-1}b + \binom{n}{2}a^{n-2}b^2 + \ldots + \binom{n}{r}a^{n-r}b^r + \ldots + b^n \quad (n \in \mathbb{N})$$

where $\binom{n}{r} = {}^nC_r = \dfrac{n!}{r!(n-r)!}$

Logarithms and exponentials

$$\log_a x = \frac{\log_b x}{\log_b a}$$

$$e^{x \ln a} = a^x$$

Differentiation

$$f'(x) = \lim_{h \to 0} \frac{f(x + h) - f(x)}{h}$$

Statistics

Probability

$P(A') = 1 - P(A)$

Standard deviation

Interquartile range $= \text{IQR} = Q_3 - Q_1$

For a set of n values $x_1, x_2, x_i, \ldots x_n$

$$S_{xx} = \sum (x_i - \bar{x})^2 = \sum x_i^2 - \frac{(\sum x_i)^2}{n}$$

Standard deviation $= \sqrt{\dfrac{S_{xx}}{n}}$ or $\sqrt{\dfrac{\sum x^2}{n - \bar{x}^2}}$

Mechanics

$v = u + at$

$s = ut + \dfrac{1}{2}at^2$

$s = vt - \dfrac{1}{2}at^2$

$v^2 = u^2 + 2as$

$s = \dfrac{1}{2}(u + v)t$

Binomial cumulative distribution function

The tabulated value is $P(X \leqslant x)$, where X has a binomial distribution with index n and parameter p.

$p =$	0.05	0.10	0.15	0.20	0.25	0.30	0.35	0.40	0.45	0.50
$n = 5, x = 0$	0.7738	0.5905	0.4437	0.3277	0.2373	0.1681	0.1160	0.0778	0.0503	0.0312
1	0.9774	0.9185	0.8352	0.7373	0.6328	0.5282	0.4284	0.3370	0.2562	0.1875
2	0.9988	0.9914	0.9734	0.9421	0.8965	0.8369	0.7648	0.6826	0.5931	0.5000
3	1.0000	0.9995	0.9978	0.9933	0.9844	0.9692	0.9460	0.9130	0.8688	0.8125
4	1.0000	1.0000	0.9999	0.9997	0.9990	0.9976	0.9947	0.9898	0.9815	0.9688
$n = 6, x = 0$	0.7351	0.5314	0.3771	0.2621	0.1780	0.1176	0.0754	0.0467	0.0277	0.0156
1	0.9672	0.8857	0.7765	0.6554	0.5339	0.4202	0.3191	0.2333	0.1636	0.1094
2	0.9978	0.9842	0.9527	0.9011	0.8306	0.7443	0.6471	0.5443	0.4415	0.3438
3	0.9999	0.9987	0.9941	0.9830	0.9624	0.9295	0.8826	0.8208	0.7447	0.6563
4	1.0000	0.9999	0.9996	0.9984	0.9954	0.9891	0.9777	0.9590	0.9308	0.8906
5	1.0000	1.0000	1.0000	0.9999	0.9998	0.9993	0.9982	0.9959	0.9917	0.9844
$n = 7, x = 0$	0.6983	0.4783	0.3206	0.2097	0.1335	0.0824	0.0490	0.0280	0.0152	0.0078
1	0.9556	0.8503	0.7166	0.5767	0.4449	0.3294	0.2338	0.1586	0.1024	0.0625
2	0.9962	0.9743	0.9262	0.8520	0.7564	0.6471	0.5323	0.4199	0.3164	0.2266
3	0.9998	0.9973	0.9879	0.9667	0.9294	0.8740	0.8002	0.7102	0.6083	0.5000
4	1.0000	0.9998	0.9988	0.9953	0.9871	0.9712	0.9444	0.9037	0.8471	0.7734
5	1.0000	1.0000	0.9999	0.9996	0.9987	0.9962	0.9910	0.9812	0.9643	0.9375
6	1.0000	1.0000	1.0000	1.0000	0.9999	0.9998	0.9994	0.9984	0.9963	0.9922
$n = 8, x = 0$	0.6634	0.4305	0.2725	0.1678	0.1001	0.0576	0.0319	0.0168	0.0084	0.0039
1	0.9428	0.8131	0.6572	0.5033	0.3671	0.2553	0.1691	0.1064	0.0632	0.0352
2	0.9942	0.9619	0.8948	0.7969	0.6785	0.5518	0.4278	0.3154	0.2201	0.1445
3	0.9996	0.9950	0.9786	0.9437	0.8862	0.8059	0.7064	0.5941	0.4770	0.3633
4	1.0000	0.9996	0.9971	0.9896	0.9727	0.9420	0.8939	0.8263	0.7396	0.6367
5	1.0000	1.0000	0.9998	0.9988	0.9958	0.9887	0.9747	0.9502	0.9115	0.8555
6	1.0000	1.0000	1.0000	0.9999	0.9996	0.9987	0.9964	0.9915	0.9819	0.9648
7	1.0000	1.0000	1.0000	1.0000	1.0000	0.9999	0.9998	0.9993	0.9983	0.9961
$n = 9, x = 0$	0.6302	0.3874	0.2316	0.1342	0.0751	0.0404	0.0207	0.0101	0.0046	0.0020
1	0.9288	0.7748	0.5995	0.4362	0.3003	0.1960	0.1211	0.0705	0.0385	0.0195
2	0.9916	0.9470	0.8591	0.7382	0.6007	0.4628	0.3373	0.2318	0.1495	0.0898
3	0.9994	0.9917	0.9661	0.9144	0.8343	0.7297	0.6089	0.4826	0.3614	0.2539
4	1.0000	0.9991	0.9944	0.9804	0.9511	0.9012	0.8283	0.7334	0.6214	0.5000
5	1.0000	0.9999	0.9994	0.9969	0.9900	0.9747	0.9464	0.9006	0.8342	0.7461
6	1.0000	1.0000	1.0000	0.9997	0.9987	0.9957	0.9888	0.9750	0.9502	0.9102
7	1.0000	1.0000	1.0000	1.0000	0.9999	0.9996	0.9986	0.9962	0.9909	0.9805
8	1.0000	1.0000	1.0000	1.0000	1.0000	1.0000	0.9999	0.9997	0.9992	0.9980
$n = 10, x = 0$	0.5987	0.3487	0.1969	0.1074	0.0563	0.0282	0.0135	0.0060	0.0025	0.0010
1	0.9139	0.7361	0.5443	0.3758	0.2440	0.1493	0.0860	0.0464	0.0233	0.0107
2	0.9885	0.9298	0.8202	0.6778	0.5256	0.3828	0.2616	0.1673	0.0996	0.0547
3	0.9990	0.9872	0.9500	0.8791	0.7759	0.6496	0.5138	0.3823	0.2660	0.1719
4	0.9999	0.9984	0.9901	0.9672	0.9219	0.8497	0.7515	0.6331	0.5044	0.3770
5	1.0000	0.9999	0.9986	0.9936	0.9803	0.9527	0.9051	0.8338	0.7384	0.6230
6	1.0000	1.0000	0.9999	0.9991	0.9965	0.9894	0.9740	0.9452	0.8980	0.8281
7	1.0000	1.0000	1.0000	0.9999	0.9996	0.9984	0.9952	0.9877	0.9726	0.9453
8	1.0000	1.0000	1.0000	1.0000	1.0000	0.9999	0.9995	0.9983	0.9955	0.9893
9	1.0000	1.0000	1.0000	1.0000	1.0000	1.0000	1.0000	0.9999	0.9997	0.9990

$p =$	0.05	0.10	0.15	0.20	0.25	0.30	0.35	0.40	0.45	0.50
$n = 12, x = 0$	0.5404	0.2824	0.1422	0.0687	0.0317	0.0138	0.0057	0.0022	0.0008	0.0002
1	0.8816	0.6590	0.4435	0.2749	0.1584	0.0850	0.0424	0.0196	0.0083	0.0032
2	0.9804	0.8891	0.7358	0.5583	0.3907	0.2528	0.1513	0.0834	0.0421	0.0193
3	0.9978	0.9744	0.9078	0.7946	0.6488	0.4925	0.3467	0.2253	0.1345	0.0730
4	0.9998	0.9957	0.9761	0.9274	0.8424	0.7237	0.5833	0.4382	0.3044	0.1938
5	1.0000	0.9995	0.9954	0.9806	0.9456	0.8822	0.7873	0.6652	0.5269	0.3872
6	1.0000	0.9999	0.9993	0.9961	0.9857	0.9614	0.9154	0.8418	0.7393	0.6128
7	1.0000	1.0000	0.9999	0.9994	0.9972	0.9905	0.9745	0.9427	0.8883	0.8062
8	1.0000	1.0000	1.0000	0.9999	0.9996	0.9983	0.9944	0.9847	0.9644	0.9270
9	1.0000	1.0000	1.0000	1.0000	1.0000	0.9998	0.9992	0.9972	0.9921	0.9807
10	1.0000	1.0000	1.0000	1.0000	1.0000	1.0000	0.9999	0.9997	0.9989	0.9968
11	1.0000	1.0000	1.0000	1.0000	1.0000	1.0000	1.0000	1.0000	0.9999	0.9998
$n = 15, x = 0$	0.4633	0.2059	0.0874	0.0352	0.0134	0.0047	0.0016	0.0005	0.0001	0.0000
1	0.8290	0.5490	0.3186	0.1671	0.0802	0.0353	0.0142	0.0052	0.0017	0.0005
2	0.9638	0.8159	0.6042	0.3980	0.2361	0.1268	0.0617	0.0271	0.0107	0.0037
3	0.9945	0.9444	0.8227	0.6482	0.4613	0.2969	0.1727	0.0905	0.0424	0.0176
4	0.9994	0.9873	0.9383	0.8358	0.6865	0.5155	0.3519	0.2173	0.1204	0.0592
5	0.9999	0.9978	0.9832	0.9389	0.8516	0.7216	0.5643	0.4032	0.2608	0.1509
6	1.0000	0.9997	0.9964	0.9819	0.9434	0.8689	0.7548	0.6098	0.4522	0.3036
7	1.0000	1.0000	0.9994	0.9958	0.9827	0.9500	0.8868	0.7869	0.6535	0.5000
8	1.0000	1.0000	0.9999	0.9992	0.9958	0.9848	0.9578	0.9050	0.8182	0.6964
9	1.0000	1.0000	1.0000	0.9999	0.9992	0.9963	0.9876	0.9662	0.9231	0.8491
10	1.0000	1.0000	1.0000	1.0000	0.9999	0.9993	0.9972	0.9907	0.9745	0.9408
11	1.0000	1.0000	1.0000	1.0000	1.0000	0.9999	0.9995	0.9981	0.9937	0.9824
12	1.0000	1.0000	1.0000	1.0000	1.0000	1.0000	0.9999	0.9997	0.9989	0.9963
13	1.0000	1.0000	1.0000	1.0000	1.0000	1.0000	1.0000	1.0000	0.9999	0.9995
14	1.0000	1.0000	1.0000	1.0000	1.0000	1.0000	1.0000	1.0000	1.0000	1.0000
$n = 20, x = 0$	0.3585	0.1216	0.0388	0.0115	0.0032	0.0008	0.0002	0.0000	0.0000	0.0000
1	0.7358	0.3917	0.1756	0.0692	0.0243	0.0076	0.0021	0.0005	0.0001	0.0000
2	0.9245	0.6769	0.4049	0.2061	0.0913	0.0355	0.0121	0.0036	0.0009	0.0002
3	0.9841	0.8670	0.6477	0.4114	0.2252	0.1071	0.0444	0.0160	0.0049	0.0013
4	0.9974	0.9568	0.8298	0.6296	0.4148	0.2375	0.1182	0.0510	0.0189	0.0059
5	0.9997	0.9887	0.9327	0.8042	0.6172	0.4164	0.2454	0.1256	0.0553	0.0207
6	1.0000	0.9976	0.9781	0.9133	0.7858	0.6080	0.4166	0.2500	0.1299	0.0577
7	1.0000	0.9996	0.9941	0.9679	0.8982	0.7723	0.6010	0.4159	0.2520	0.1316
8	1.0000	0.9999	0.9987	0.9900	0.9591	0.8867	0.7624	0.5956	0.4143	0.2517
9	1.0000	1.0000	0.9998	0.9974	0.9861	0.9520	0.8782	0.7553	0.5914	0.4119
10	1.0000	1.0000	1.0000	0.9994	0.9961	0.9829	0.9468	0.8725	0.7507	0.5881
11	1.0000	1.0000	1.0000	0.9999	0.9991	0.9949	0.9804	0.9435	0.8692	0.7483
12	1.0000	1.0000	1.0000	1.0000	0.9998	0.9987	0.9940	0.9790	0.9420	0.8684
13	1.0000	1.0000	1.0000	1.0000	1.0000	0.9997	0.9985	0.9935	0.9786	0.9423
14	1.0000	1.0000	1.0000	1.0000	1.0000	1.0000	0.9997	0.9984	0.9936	0.9793
15	1.0000	1.0000	1.0000	1.0000	1.0000	1.0000	1.0000	0.9997	0.9985	0.9941
16	1.0000	1.0000	1.0000	1.0000	1.0000	1.0000	1.0000	1.0000	0.9997	0.9987
17	1.0000	1.0000	1.0000	1.0000	1.0000	1.0000	1.0000	1.0000	1.0000	0.9998
18	1.0000	1.0000	1.0000	1.0000	1.0000	1.0000	1.0000	1.0000	1.0000	1.0000

$p =$	0.05	0.10	0.15	0.20	0.25	0.30	0.35	0.40	0.45	0.50
$n = 25, x = 0$	0.2774	0.0718	0.0172	0.0038	0.0008	0.0001	0.0000	0.0000	0.0000	0.0000
1	0.6424	0.2712	0.0931	0.0274	0.0070	0.0016	0.0003	0.0001	0.0000	0.0000
2	0.8729	0.5371	0.2537	0.0982	0.0321	0.0090	0.0021	0.0004	0.0001	0.0000
3	0.9659	0.7636	0.4711	0.2340	0.0962	0.0332	0.0097	0.0024	0.0005	0.0001
4	0.9928	0.9020	0.6821	0.4207	0.2137	0.0905	0.0320	0.0095	0.0023	0.0005
5	0.9988	0.9666	0.8385	0.6167	0.3783	0.1935	0.0826	0.0294	0.0086	0.0020
6	0.9998	0.9905	0.9305	0.7800	0.5611	0.3407	0.1734	0.0736	0.0258	0.0073
7	1.0000	0.9977	0.9745	0.8909	0.7265	0.5118	0.3061	0.1536	0.0639	0.0216
8	1.0000	0.9995	0.9920	0.9532	0.8506	0.6769	0.4668	0.2735	0.1340	0.0539
9	1.0000	0.9999	0.9979	0.9827	0.9287	0.8106	0.6303	0.4246	0.2424	0.1148
10	1.0000	1.0000	0.9995	0.9944	0.9703	0.9022	0.7712	0.5858	0.3843	0.2122
11	1.0000	1.0000	0.9999	0.9985	0.9893	0.9558	0.8746	0.7323	0.5426	0.3450
12	1.0000	1.0000	1.0000	0.9996	0.9966	0.9825	0.9396	0.8462	0.6937	0.5000
13	1.0000	1.0000	1.0000	0.9999	0.9991	0.9940	0.9745	0.9222	0.8173	0.6550
14	1.0000	1.0000	1.0000	1.0000	0.9998	0.9982	0.9907	0.9656	0.9040	0.7878
15	1.0000	1.0000	1.0000	1.0000	1.0000	0.9995	0.9971	0.9868	0.9560	0.8852
16	1.0000	1.0000	1.0000	1.0000	1.0000	0.9999	0.9992	0.9957	0.9826	0.9461
17	1.0000	1.0000	1.0000	1.0000	1.0000	1.0000	0.9998	0.9988	0.9942	0.9784
18	1.0000	1.0000	1.0000	1.0000	1.0000	1.0000	1.0000	0.9997	0.9984	0.9927
19	1.0000	1.0000	1.0000	1.0000	1.0000	1.0000	1.0000	0.9999	0.9996	0.9980
20	1.0000	1.0000	1.0000	1.0000	1.0000	1.0000	1.0000	1.0000	0.9999	0.9995
21	1.0000	1.0000	1.0000	1.0000	1.0000	1.0000	1.0000	1.0000	1.0000	0.9999
22	1.0000	1.0000	1.0000	1.0000	1.0000	1.0000	1.0000	1.0000	1.0000	1.0000
$n = 30, x = 0$	0.2146	0.0424	0.0076	0.0012	0.0002	0.0000	0.0000	0.0000	0.0000	0.0000
1	0.5535	0.1837	0.0480	0.0105	0.0020	0.0003	0.0000	0.0000	0.0000	0.0000
2	0.8122	0.4114	0.1514	0.0442	0.0106	0.0021	0.0003	0.0000	0.0000	0.0000
3	0.9392	0.6474	0.3217	0.1227	0.0374	0.0093	0.0019	0.0003	0.0000	0.0000
4	0.9844	0.8245	0.5245	0.2552	0.0979	0.0302	0.0075	0.0015	0.0002	0.0000
5	0.9967	0.9268	0.7106	0.4275	0.2026	0.0766	0.0233	0.0057	0.0011	0.0002
6	0.9994	0.9742	0.8474	0.6070	0.3481	0.1595	0.0586	0.0172	0.0040	0.0007
7	0.9999	0.9922	0.9302	0.7608	0.5143	0.2814	0.1238	0.0435	0.0121	0.0026
8	1.0000	0.9980	0.9722	0.8713	0.6736	0.4315	0.2247	0.0940	0.0312	0.0081
9	1.0000	0.9995	0.9903	0.9389	0.8034	0.5888	0.3575	0.1763	0.0694	0.0214
10	1.0000	0.9999	0.9971	0.9744	0.8943	0.7304	0.5078	0.2915	0.1350	0.0494
11	1.0000	1.0000	0.9992	0.9905	0.9493	0.8407	0.6548	0.4311	0.2327	0.1002
12	1.0000	1.0000	0.9998	0.9969	0.9784	0.9155	0.7802	0.5785	0.3592	0.1808
13	1.0000	1.0000	1.0000	0.9991	0.9918	0.9599	0.8737	0.7145	0.5025	0.2923
14	1.0000	1.0000	1.0000	0.9998	0.9973	0.9831	0.9348	0.8246	0.6448	0.4278
15	1.0000	1.0000	1.0000	0.9999	0.9992	0.9936	0.9699	0.9029	0.7691	0.5722
16	1.0000	1.0000	1.0000	1.0000	0.9998	0.9979	0.9876	0.9519	0.8644	0.7077
17	1.0000	1.0000	1.0000	1.0000	0.9999	0.9994	0.9955	0.9788	0.9286	0.8192
18	1.0000	1.0000	1.0000	1.0000	1.0000	0.9998	0.9986	0.9917	0.9666	0.8998
19	1.0000	1.0000	1.0000	1.0000	1.0000	1.0000	0.9996	0.9971	0.9862	0.9506
20	1.0000	1.0000	1.0000	1.0000	1.0000	1.0000	0.9999	0.9991	0.9950	0.9786
21	1.0000	1.0000	1.0000	1.0000	1.0000	1.0000	1.0000	0.9998	0.9984	0.9919
22	1.0000	1.0000	1.0000	1.0000	1.0000	1.0000	1.0000	1.0000	0.9996	0.9974
23	1.0000	1.0000	1.0000	1.0000	1.0000	1.0000	1.0000	1.0000	0.9999	0.9993
24	1.0000	1.0000	1.0000	1.0000	1.0000	1.0000	1.0000	1.0000	1.0000	0.9998
25	1.0000	1.0000	1.0000	1.0000	1.0000	1.0000	1.0000	1.0000	1.0000	1.0000

p =	0.05	0.10	0.15	0.20	0.25	0.30	0.35	0.40	0.45	0.50
$n = 40, x = 0$	0.1285	0.0148	0.0015	0.0001	0.0000	0.0000	0.0000	0.0000	0.0000	0.0000
1	0.3991	0.0805	0.0121	0.0015	0.0001	0.0000	0.0000	0.0000	0.0000	0.0000
2	0.6767	0.2228	0.0486	0.0079	0.0010	0.0001	0.0000	0.0000	0.0000	0.0000
3	0.8619	0.4231	0.1302	0.0285	0.0047	0.0006	0.0001	0.0000	0.0000	0.0000
4	0.9520	0.6290	0.2633	0.0759	0.0160	0.0026	0.0003	0.0000	0.0000	0.0000
5	0.9861	0.7937	0.4325	0.1613	0.0433	0.0086	0.0013	0.0001	0.0000	0.0000
6	0.9966	0.9005	0.6067	0.2859	0.0962	0.0238	0.0044	0.0006	0.0001	0.0000
7	0.9993	0.9581	0.7559	0.4371	0.1820	0.0553	0.0124	0.0021	0.0002	0.0000
8	0.9999	0.9845	0.8646	0.5931	0.2998	0.1110	0.0303	0.0061	0.0009	0.0001
9	1.0000	0.9949	0.9328	0.7318	0.4395	0.1959	0.0644	0.0156	0.0027	0.0003
10	1.0000	0.9985	0.9701	0.8392	0.5839	0.3087	0.1215	0.0352	0.0074	0.0011
11	1.0000	0.9996	0.9880	0.9125	0.7151	0.4406	0.2053	0.0709	0.0179	0.0032
12	1.0000	0.9999	0.9957	0.9568	0.8209	0.5772	0.3143	0.1285	0.0386	0.0083
13	1.0000	1.0000	0.9986	0.9806	0.8968	0.7032	0.4408	0.2112	0.0751	0.0192
14	1.0000	1.0000	0.9996	0.9921	0.9456	0.8074	0.5721	0.3174	0.1326	0.0403
15	1.0000	1.0000	0.9999	0.9971	0.9738	0.8849	0.6946	0.4402	0.2142	0.0769
16	1.0000	1.0000	1.0000	0.9990	0.9884	0.9367	0.7978	0.5681	0.3185	0.1341
17	1.0000	1.0000	1.0000	0.9997	0.9953	0.9680	0.8761	0.6885	0.4391	0.2148
18	1.0000	1.0000	1.0000	0.9999	0.9983	0.9852	0.9301	0.7911	0.5651	0.3179
19	1.0000	1.0000	1.0000	1.0000	0.9994	0.9937	0.9637	0.8702	0.6844	0.4373
20	1.0000	1.0000	1.0000	1.0000	0.9998	0.9976	0.9827	0.9256	0.7870	0.5627
21	1.0000	1.0000	1.0000	1.0000	1.0000	0.9991	0.9925	0.9608	0.8669	0.6821
22	1.0000	1.0000	1.0000	1.0000	1.0000	0.9997	0.9970	0.9811	0.9233	0.7852
23	1.0000	1.0000	1.0000	1.0000	1.0000	0.9999	0.9989	0.9917	0.9595	0.8659
24	1.0000	1.0000	1.0000	1.0000	1.0000	1.0000	0.9996	0.9966	0.9804	0.9231
25	1.0000	1.0000	1.0000	1.0000	1.0000	1.0000	0.9999	0.9988	0.9914	0.9597
26	1.0000	1.0000	1.0000	1.0000	1.0000	1.0000	1.0000	0.9996	0.9966	0.9808
27	1.0000	1.0000	1.0000	1.0000	1.0000	1.0000	1.0000	0.9999	0.9988	0.9917
28	1.0000	1.0000	1.0000	1.0000	1.0000	1.0000	1.0000	1.0000	0.9996	0.9968
29	1.0000	1.0000	1.0000	1.0000	1.0000	1.0000	1.0000	1.0000	0.9999	0.9989
30	1.0000	1.0000	1.0000	1.0000	1.0000	1.0000	1.0000	1.0000	1.0000	0.9997
31	1.0000	1.0000	1.0000	1.0000	1.0000	1.0000	1.0000	1.0000	1.0000	0.9999
32	1.0000	1.0000	1.0000	1.0000	1.0000	1.0000	1.0000	1.0000	1.0000	1.0000

$p =$	0.05	0.10	0.15	0.20	0.25	0.30	0.35	0.40	0.45	0.50
$n = 50, x = 0$	0.0769	0.0052	0.0003	0.0000	0.0000	0.0000	0.0000	0.0000	0.0000	0.0000
1	0.2794	0.0338	0.0029	0.0002	0.0000	0.0000	0.0000	0.0000	0.0000	0.0000
2	0.5405	0.1117	0.0142	0.0013	0.0001	0.0000	0.0000	0.0000	0.0000	0.0000
3	0.7604	0.2503	0.0460	0.0057	0.0005	0.0000	0.0000	0.0000	0.0000	0.0000
4	0.8964	0.4312	0.1121	0.0185	0.0021	0.0002	0.0000	0.0000	0.0000	0.0000
5	0.9622	0.6161	0.2194	0.0480	0.0070	0.0007	0.0001	0.0000	0.0000	0.0000
6	0.9882	0.7702	0.3613	0.1034	0.0194	0.0025	0.0002	0.0000	0.0000	0.0000
7	0.9968	0.8779	0.5188	0.1904	0.0453	0.0073	0.0008	0.0001	0.0000	0.0000
8	0.9992	0.9421	0.6681	0.3073	0.0916	0.0183	0.0025	0.0002	0.0000	0.0000
9	0.9998	0.9755	0.7911	0.4437	0.1637	0.0402	0.0067	0.0008	0.0001	0.0000
10	1.0000	0.9906	0.8801	0.5836	0.2622	0.0789	0.0160	0.0022	0.0002	0.0000
11	1.0000	0.9968	0.9372	0.7107	0.3816	0.1390	0.0342	0.0057	0.0006	0.0000
12	1.0000	0.9990	0.9699	0.8139	0.5110	0.2229	0.0661	0.0133	0.0018	0.0002
13	1.0000	0.9997	0.9868	0.8894	0.6370	0.3279	0.1163	0.0280	0.0045	0.0005
14	1.0000	0.9999	0.9947	0.9393	0.7481	0.4468	0.1878	0.0540	0.0104	0.0013
15	1.0000	1.0000	0.9981	0.9692	0.8369	0.5692	0.2801	0.0955	0.0220	0.0033
16	1.0000	1.0000	0.9993	0.9856	0.9017	0.6839	0.3889	0.1561	0.0427	0.0077
17	1.0000	1.0000	0.9998	0.9937	0.9449	0.7822	0.5060	0.2369	0.0765	0.0164
18	1.0000	1.0000	0.9999	0.9975	0.9713	0.8594	0.6216	0.3356	0.1273	0.0325
19	1.0000	1.0000	1.0000	0.9991	0.9861	0.9152	0.7264	0.4465	0.1974	0.0595
20	1.0000	1.0000	1.0000	0.9997	0.9937	0.9522	0.8139	0.5610	0.2862	0.1013
21	1.0000	1.0000	1.0000	0.9999	0.9974	0.9749	0.8813	0.6701	0.3900	0.1611
22	1.0000	1.0000	1.0000	1.0000	0.9990	0.9877	0.9290	0.7660	0.5019	0.2399
23	1.0000	1.0000	1.0000	1.0000	0.9996	0.9944	0.9604	0.8438	0.6134	0.3359
24	1.0000	1.0000	1.0000	1.0000	0.9999	0.9976	0.9793	0.9022	0.7160	0.4439
25	1.0000	1.0000	1.0000	1.0000	1.0000	0.9991	0.9900	0.9427	0.8034	0.5561
26	1.0000	1.0000	1.0000	1.0000	1.0000	0.9997	0.9955	0.9686	0.8721	0.6641
27	1.0000	1.0000	1.0000	1.0000	1.0000	0.9999	0.9981	0.9840	0.9220	0.7601
28	1.0000	1.0000	1.0000	1.0000	1.0000	1.0000	0.9993	0.9924	0.9556	0.8389
29	1.0000	1.0000	1.0000	1.0000	1.0000	1.0000	0.9997	0.9966	0.9765	0.8987
30	1.0000	1.0000	1.0000	1.0000	1.0000	1.0000	0.9999	0.9986	0.9884	0.9405
31	1.0000	1.0000	1.0000	1.0000	1.0000	1.0000	1.0000	0.9995	0.9947	0.9675
32	1.0000	1.0000	1.0000	1.0000	1.0000	1.0000	1.0000	0.9998	0.9978	0.9836
33	1.0000	1.0000	1.0000	1.0000	1.0000	1.0000	1.0000	0.9999	0.9991	0.9923
34	1.0000	1.0000	1.0000	1.0000	1.0000	1.0000	1.0000	1.0000	0.9997	0.9967
35	1.0000	1.0000	1.0000	1.0000	1.0000	1.0000	1.0000	1.0000	0.9999	0.9987
36	1.0000	1.0000	1.0000	1.0000	1.0000	1.0000	1.0000	1.0000	1.0000	0.9995
37	1.0000	1.0000	1.0000	1.0000	1.0000	1.0000	1.0000	1.0000	1.0000	0.9998
38	1.0000	1.0000	1.0000	1.0000	1.0000	1.0000	1.0000	1.0000	1.0000	1.0000